1989

Cargo Access Equipment for Merchant Ships

Cargo Access Equipment for Merchant Ships

I. L. BUXTON
Reader in Marine Transport
Department of Naval Architecture and Shipbuilding
University of Newcastle-upon-Tyne

R. P. DAGGITT
Naval Architect, Sandock Austral, Durban
Formerly Research Associate, Department of Naval Architecture and Shipbuilding
University of Newcastle-upon-Tyne

J. KING
Professor of Maritime Technology
University of Wales Institute of Science and Technology
Formerly Lecturer, Department of Naval Architecture and Shipbuilding
University of Newcastle-upon-Tyne

LONDON
E. & F. N. SPON LIMITED

First published 1978
by E. & F. N. Spon Limited
11 New Fetter Lane, London EC4P 4EE

© *1978 MacGregor Publications Ltd.*

Printed in Great Britain by
Richard Clay (The Chaucer Press) Ltd.
Bungay, Suffolk

ISBN 0 419 11490 4

Foreword

As President of International MacGregor I am deeply indebted to the authors of this excellent book for the very considerable amount of work and scholarship it contains. It is the first authoritative work on cargo access equipment to be published and I am sure that it will be greatly welcomed by the Marine Industries.

You will see from the authors' preface that the book was commissioned by the Henri Kummerman Foundation which was established in 1976 to assist and promote internationally research and development in the field of marine transportation and cargo handling.

The Foundation has already made a number of grants to universities and to students but this book is its first major contribution to the furthering of education in the Marine Industries. For me, it is a rewarding fruition of a long involvement in maritime affairs.

However, much requires to be done in the future and the Foundation can only succeed if it is encouraged and assisted by people who are forward thinking. I should be pleased therefore to hear from any readers of this book if they feel that they can help or be helped within the aims and objectives of the Foundation.

28 Chemin du Pommier,   HENRI KUMMERMAN
1218 Geneva,
Switzerland.
May 1978

Contents

Contents

Preface

Shipboard cargo access equipment includes hatch covers, bow, stern and side doors, ramps, elevators and movable decks, all of which facilitate loading and discharging operations. Such equipment can account for 10% of a ship's initial cost. Worldwide some $400 million is spent annually on equipping over 1000 ships. Yet, despite this importance, enhanced by the rapidly developing field of roll-on/roll-off (Ro-Ro) shipping, no comprehensive publication has ever been produced on cargo access equipment.

Recognizing the need for such a book, the President of the International MacGregor Organization, Henri Kummerman, through his Foundation commissioned the School of Marine Technology at the University of Newcastle upon Tyne to prepare a definitive reference work.

The MacGregor Organization is the world's largest manufacturer of cargo access equipment, and so widespread is the use of its products that the description 'MacGregor hatch' is often loosely used to describe any type of mechanical steel hatch cover.

In this book we have tried to cover all aspects of cargo access, including design requirements, description of principal types of equipment, economics of selection and operation, and performance in service. For those already familiar with cargo access equipment, we have included useful reference material such as weights and stowage requirements; while for those new to the subject, we have included sufficient background material on ship types and cargo handling for developments such as Ro-Ro shipping to be fully appreciated. The book does not confine itself to MacGregor products, but draws its material from all over the world and includes examples of the work of other important companies in the field such as Navire Cargo Gear and Kvaerner Brug.

The authors are not specialists in the field of cargo access equipment, but draw their experience from ship design, ship operation and mechanical engineering backgrounds. We have therefore greatly appreciated the co-operation of staff within the MacGregor Organization, especially those in the London, Whitley Bay and Paris offices, who have provided much of the basic information and expert guidance essential to such an undertaking. We are also grateful to the many people in the shipowning, shipbuilding, ship repairing, port operations and cargo handling equipment industries who have helped us, as indicated in the acknowledgements.

We hope that the book will be useful not only to practitioners in the marine industries but also to institutions concerned with the teaching of marine technology – universities, polytechnics and nautical colleges – and also to the staffs of marine equipment companies, both as an introduction for newcomers and as a reference book for the more experienced.

We pondered the most suitable measurement units to use – SI, metric or imperial – but felt that the book's value to practical people around the world would be increased by the general use of metric units, e.g. for stresses, but with occasional use of imperial units where these have a special significance, e.g. for container sizes, or when they are still commonly employed, e.g. for drafts or stowage factors.

School of Marine Technology   I.L.B.
University of Newcastle upon Tyne   R.P.D.
November 1977   J.K.

Acknowledgements

The production of a book such as this necessarily involves the assistance and co-operation of many individuals and organizations. We are particularly indebted to Mr P. W. Penney, lecturer in the Department of Naval Architecture and Shipbuilding in the University of Newcastle upon Tyne for contributing Chapter 11 and to the departmental secretarial staff for never failing to turn our illegible notes into clear and accurate typescript. We should also like to record our thanks to the organizations listed below.

American Bureau of Shipping
Atlantic Container Lines Services
Austin & Pickersgill
Australind S.S. Co.
Australian High Commission, London
British Rail Shipping and International Services Division
British Road Services
British Steel Corporation
British Transport Docks Board
Bureau Veritas
Cammell Laird Shipbuilders
Cargospeed Equipment
Common Brothers
Crane Fruehauf Containers
Department of Trade, Marine Division
D.F.D.S., Copenhagen and North Shields
European Ferries
Germanischer Lloyd
International Cargo Handling Coordination Association
Kvaerner Brug
Lancer Boss
Lansing Henley
J. Lauritzen
Lloyds Register of Shipping, London and Newcastle
MAFI Technical Systems
Marine Transport Centre, University of Liverpool

MacGregor Organization offices: United Kingdom, France, Holland, Germany,
   Japan, USA
Thomas R. Miller & Son
National Ports Council
Navire Cargo Gear
New South Wales Department of Transport
Norske Veritas
Overseas Containers Limited
Ocean Fleets
P & O Ferries
P & O Strath Services
Port of London Authority
Royal Institution of Naval Architects
Salen and Wicander
Sea Containers Services
Seasafe Transport
Seaspeed Ferries
Sun Shipbuilding and Dry Dock
Swan Hunter Shipbuilders
Swan Hunter Shiprepairers
Swedish Road Safety Office
Tor Line
Valmet Oy
York Trailer

Many of the illustrations used in the text have been supplied either by the MacGregor Organization or by the authors, but acknowledgements are due to the following on whose publications we have drawn freely.

Atlantic Container Lines: Fig. 2.15; Blohm and Voss: Fig. 6.56; British Transport Docks Board: Figs. 2.4, 6.16; Business Meetings Limited (*Ro-Ro 76*): Fig. 11.1 Cargo Systems International: Fig. 11.2; Japan Ship Exporters Association: Figs 2.13, 2.14, 8.16; Kvaerner Brug: Fig. 5.23; Lansing Henley: Fig. 6.5; Lloyd's Register of Shipping: Figs. 9.4, 9.6; Miller Freeman Publications, 'Ship Design for Forest Products', by R. N. Herbert, from *Transport and Handling in the Pulp and Paper Industry*, Vol. 1: Fig. 7.4; *The Motor Ship*: Figs 7.2, 8.1, 8.8, 8.18, 8.20; Navire Cargo Gear: Figs 5.22(b), 5.41, 6.22, 8.17; Overseas Containers Limited: Fig. 8.7; P & O: Fig. 5.4; Port of London Authority: Fig. 8.5; Royal Institution of Naval Architects: Figs. 1.4, 1.6, 1.10, 1.11, 1.12, 1.13, 1.14, 11.5; *Shipping World and Shipbuilder*: Figs 1.3, 1.8, 1.9, 2.9, 8.9, 8.13; *Shipbuilding and Marine Engineering International*: Figs 2.17, 8.14; Skyfotos: Fig. 2.3; Valmet Oy: Fig. 6.6.

The authors' thanks are also due to Ms Roma Aldrich who has drawn the line diagrams specially commissioned for this book.

Notwithstanding the advice, comments and other help that the authors have received, the responsibility for any errors or omissions remains theirs alone.

ILB, RPD, JK

CHAPTER ONE

The Early Development of Hatch Covers

1.1 Introduction to the access problem

There is a Chinese proverb which runs: 'Water can both float and sink a ship'; a literal reading of which encapsulates one of the central dilemmas faced by naval architects through the ages. In order to comply with the laws of hydrostatics, a ship must be arranged so that its hull is buoyant and keeps water out. Yet if it is to be more than a simple raft, a ship must be arranged to allow access to its interior. To be buoyant in all the conditions that it is likely to meet in service, the ideal hull would be a hollow, watertight chamber, yet to provide access it is necessary to destroy its watertight integrity by cutting openings in its shell. Throughout maritime history many solutions to this problem have been devised, but the number of fundamentally different ways of providing access to a ship is limited to three.

If a ship is regarded as a hollow rectangular box, access may be gained horizontally through the sides, horizontally through the ends or vertically through top and bottom. These are the only alternatives and they have all been used at one time or another, either singly or in combination.

Throughout history, vertical access through the top has been by far the most common. Many early craft were small open boats for which there was, of course, no practical alternative, but there are records of ancient craft that were decked over and fitted with hatchways. To the present day, vertical access through upper deck hatches remains the usual arrangement for most dry cargo ships.

Vertical access through the bottom has been much less common. This is hardly surprising. Nonetheless it has been adopted for certain vessels, notably hopper barges for transporting dredged material and dumping it over spoil grounds by the simple expedient of opening bottom doors. Bottom access has also been extensively used in certain bulk carriers, especially those employed on the Great Lakes. Their hopper-shaped holds have doors opening onto a conveyor running longitudinally between the inner and outer bottoms. During discharging operations, cargo is allowed to fall through the doors onto the conveyors which transport it to one end of the ship where it is transferred to a vertical hoist, lifted to the upper deck and thence ashore.

All cargo transfer processes involve a combination of horizontal and vertical movements. Nowadays horizontal movements are often associated with roll-on/roll-off operations using ramps and other contrivances whose origins can be traced to the simple gangplank which has been employed since ancient times for pedestrian access to ships. Walk-on/walk-off methods of cargo handling are still practised in some ports. However, modern roll-on/roll-off operations are distinguished from earlier methods of horizontal loading by ramps which offer access to the interior of the ship rather than to the weather deck alone. Openings in the hull must be large enough to permit easy access; they must be watertight and they must not excessively weaken the hull structure. All of these are clues to why side shell openings were not usual in ships in ancient times.

Gun-ports found in 17th and 18th century warships are the ancestors of the modern side shell door. These were small rectangular openings in the hull, each fitted with a substantial cover which was opened at sea to allow guns to bear on an enemy. They were difficult to keep watertight and were probably responsible for the loss of numerous vessels. Side shell openings were uncommon in merchant ships until late in the 19th century when they were occasionally fitted at the embarkation decks of passenger liners. Even today, such doors are often referred to as 'gun-port doors'. There are now many vessels, notably pallet ships and vehicle carriers, which are arranged for horizontal side loading.

Train ferries were among the first purpose-built roll-on/roll-off ships (now usually abbreviated to Ro-Ro) and were in use on the Rivers Forth and Tay as long ago as 1860. Rolling stock was transferred from quay to ship by means of adjustable bow and/or stern ramps in a similar manner to that which is still in common use today. Some of the earliest train ferries were even arranged so that loading and discharging could be carried out at opposite ends – a feature of many of the latest Ro-Ro ferries also.

1.2 Early hatch covers

The upper deck hatches of early steamships, in common with those of sailing ships, were small by present standards; four metres by two would not have been untypical hatchway dimensions, even for ocean-going vessels in the mid-19th century. Deck openings were bounded by vertical coamings and spanned by wooden boards usually laid athwartships; watertightness was maintained by tarpaulins spread over the boards and secured by wedges and cleats at the coamings, an arrangement which had then been in use for centuries. Web beams were sometimes placed across the largest hatchways but it was not until 1879 that their use became mandatory for all vessels classed with Lloyd's Register of Shipping having hatchways more than 12 ft (3·7 m) long. Coamings were required to be of iron construction. No minimum height was specified and it was not until nearly forty years later that they appeared in Lloyd's Rules. Hatch boards were made of solid wood $2\frac{1}{2}$ in (63 mm) thick and of such an overall size that they could be handled manually. Since they were laid transversely across the hatchway it was necessary to use 'fore and afters' to reduce their unsupported span. These were heavy baulks of timber placed longitudinally across the hatchway, on top of the transverse iron beams and supported at each end by brackets attached to the coamings. The hatch boards, in turn, were placed across the fore and afters which were so spaced that the unsupported span of the boards did not exceed the maximum allowed (at that time 2 ft $7\frac{1}{2}$ in (0·80 m), later increased to 5 ft (1·5 m)).

After the turn of the 20th century, the practice of arranging hatch boards longitudinally began to be adopted. There were sound reasons for this change. A typical cargo ship hatch, 7·5 m (25 ft) long and 5 m (16 ft) wide, with ath-

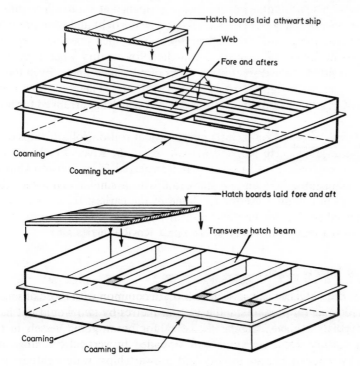

Fig. 1.1. Simplified arrangement of a traditional hatch: (*upper*) with transverse boards and fore and afters; (*lower*) with longitudinal boards laid on transverse beams.

wartship hatch boards, would have been fitted with two transverse web beams and three fore and afters which, because of their overall length, would each have been made up in three pieces, one for every beam space. With longitudinal hatch boards the same hatch would have required no more than four transverse beams spaced 1·5 m (5 ft) apart, each of lighter construction than the two webs in the previous arrangement. Thus the number of heavy components to be lifted every time the hatch was opened or closed would have been reduced from eleven to four (see Fig. 1.1).

This arrangement also had the advantage of being safer than the earlier one. As hatches became longer, it became increasingly necessary to fit fore and afters in several sections. These were more difficult to ship when

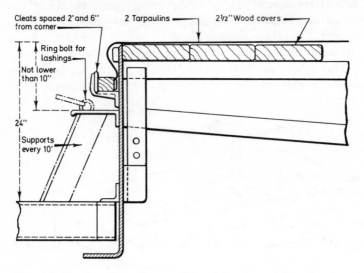

Fig. 1.2. Section through a traditional hatch coaming showing the beam landing and tarpaulin securing arrangements.

battening down than those made in one piece, since at least one end of each section had to be positioned on a seat attached to a transverse web before the hatch boards could be put on. This meant that a man had to sit astride the web, over the open hold, to guide it into place – a very dangerous practice.

Athwartship hatch boards and fore and afters continued to be used well into the 20th century although the alternative longitudinal arrangement gradually replaced them. After the implementation of the 1930 International Load Line Convention, fore and afters were only found in ships with very small hatches.

All types of wooden hatch were made watertight by means of tarpaulins and this method is still employed in those ships with traditional hatches that remain in service. Hatch tarpaulins were usually made by sewing together panels of heavy waterproofed canvas. The panels were each the width of a single bolt of canvas and were sewn together (with seams that would shed water) to form a sheet large enough to cover the entire hatchway with a generous overlap at the coamings. Hatches were usually provided with two tarpaulins, sometimes more. When more than two were available it was usual for the newest to be underneath with at least one older one on top. This practice was adopted because tarpaulins were very easily damaged by chafing, especially when locking bars (portable bars laid athwartships across the tarpaulin) or lashings were in use, and the newest tarpaulin, the primary watertight barrier, was afforded some protection by placing it underneath an older one. At the coamings the layers of canvas were folded together and slipped into cleats attached either to the vertical coaming plating or to a horizontal stiffener fitted all round the outer side of the coaming. The folds of canvas were then held in place in the cleats by means of long steel battening bars which were themselves secured by means of wooden wedges, pressing the canvas against the coaming (see Fig. 1.2).

The use of wooden hatch covers has declined throughout much of this century to the extent that it is now almost unthinkable that they should be fitted in a modern vessel of any size. It is interesting, therefore, to consider the reasons for their demise. These fall into three broad categories, namely safety and security, cargo working and maintenance.

1.3 Failings of wooden hatches

1.3.1 Safety and security

It is impossible to exaggerate the importance of efficient and secure hatch covers; there can be no question that a ship's survival may depend upon having them. But how efficient are wooden hatch covers?

Experience has shown that tarpaulins are the most vulnerable components of traditional hatch covers. There are several common reasons for tarpaulin failure. Slackening of wedges is undoubtedly an important factor. Chafing is another serious source of weakness, especially at the coamings, which has led, directly or indirectly, to the loss of innumerable vessels. Many shipowners and masters in the past were reluctant to use locking bars or lashings to provide additional security for the hatches (and to prevent the ballooning of the tarpaulins with every gust of wind down the ventilators) because of the chafing damage that they caused. Tarpaulins are also liable to suffer damage every time that they are removed for working cargo.

With the tarpaulin gone, there is nothing to prevent water entering a hold since the boards are not watertight and are easily blown off by the wind or washed away by breaking waves. In many of the accounts given by survivors of ships lost when water was taken on board, the hatches were described as having been 'stove in'. Presumably this means that they collapsed under the weight of water and in a few instances there can be no doubt that this is what happened. But so long as they are maintained in good order, and annual inspections now ensure this, wooden boards are adequately strong in relation to all the other hatch components. Thus many of the reports of hatches being stove in must be viewed with scepticism, even though they have often been accepted by the Courts. It seems more probable that the events described by witnesses began with the tearing of tarpaulins followed by the unshipping of hatch boards, rather than with the collapse of a hatch under the sheer weight of water on top of it.

During the 1920s doubts about the security of wooden hatches also arose as a result of the comparatively large size of the hatchways of many of the ships, especially colliers, then entering service. Not only were the hatchways longer and wider than had been the custom up to that time, but the ratio of hatch width to overall beam was often greater too. This was particularly apparent in ships employed in the carriage of bulk cargoes and some of the most extreme examples were to be found amongst self-trimming colliers where hatch width/ship beam ratios often exceeded 0·6.

Some observers took the view that only by the introduction of some form of steel hatch could the safety of these ships be assured, and by 1930 several types of steel hatches were available. This view was not widely held among the shipping fraternity. On the contrary, there was a great deal of opposition to the introduction of steel hatch covers and a variety of modifications and improvements were devised to overcome the evident short-comings of wooden hatches.

One of the most vociferous critics of steel covers was Harry Cocks, a Cardiff man whose patent reinforced wooden hatch board, consisting of two planks side by side with a steel bar running between them and having their ends inserted in steel shoes (see Fig. 1.3), proved so popular that many ship-owners were probably persuaded to persevere with wooden hatches because

Fig. 1.3. Reinforced wooden hatch board. Steel bands to prevent end-splitting were introduced in response to criticism of the strength and serviceability of wooden boards.

of it. Cocks' invention, like so many which have lasting influence, was extremely simple yet of such obvious benefit that one wonders that it was so long in coming. Wooden hatch boards suffered heavy wear and tear in use; their (usually) unprotected ends were easily damaged with the result that they eventually became weakened and split. Reinforced hatch boards were not only stronger than those of orthodox pattern, they were also more reliable. Equally important perhaps, they did not require any change in traditional practices.

The mechanical simplicity of boards, beams and tarpaulins was seen as a considerable advantage by shipowners and seamen alike, and this accounts for the popularity during the 1930s of the Isherwood hatch cover. Like Cocks, Isherwood, whose name is associated with several important developments in naval architecture, identified the hatch board as a principal source of weakness, but his solution was to substitute easily manhandled steel panels for wooden boards. These steel panels were designed to be interchangeable with wooden boards and they could therefore be fitted in ships which had originally been constructed with wooden hatches. Their virtues were their strength and durability, but tarpaulins were still needed to keep water out.

Individual hatch boards were laid loosely on top of the beams in a normal hatchway and they could easily be unshipped accidentally. In order to prevent this, interlocking boards were developed. These could be handled manually in the usual manner but once landed on the beams, they were retained in place by a simple locking device which was claimed by its manufacturer to enhance the safety of a ship, should it lose its tarpaulins, by preventing the boards from being washed away by the sea. This claim was probably a fair one but, like so many of the developments which were introduced around this time, it contributed nothing to the cure of the fundamental weakness of the traditional hatch, which was the vulnerability of its tarpaulins.

Safety and security were not, however, the only grounds on which these hatches were found wanting. If their simplicity was in one sense a virtue, there was another important sense in which it could be regarded a major drawback.

1.3.2 Cargo working

The influence of time spent in port on the overall economics of operating ships has long been recognized. As long ago as the beginning of the present century, attention was being drawn to the need for higher cargo handling rates and faster turnround times, and time spent opening and closing hatches was identified as a significant element in the total time in port. This became increasingly apparent as hatchways became larger. Every hatch component had to be handled separately; tarpaulins had to be folded back, each hatch board had to be removed manually, each beam had to be lifted from its seat

Fig. 1.4. A typical traditional hatch. Note the arrangement of the beams and the hatch boards stowed roughly alongside the coamings.

by derrick and winch or shore crane, and the larger the hatchway, the more items there were to be handled. Some improvement was achieved as a result of the adoption of longitudinal rather than transverse hatch boards because the absence of fore and afters meant that fewer heavy components had to be lifted in and out. Further improvements were needed however, and many were proposed.

Single hatch boards, depending upon their length, could be carried quite easily by one or two men. Whenever a hatch was opened, boards had to be removed from their positions in the hatchway and carried to the deck abreast the side coamings, their customary stowage place. Stripping an entire hatch was obviously a lengthy process when carried out in this way and was not undertaken unless it was absolutely necessary. To save time in general cargo ships, for instance, the usual practice was to remove only those boards immediately above the part of the hold being worked.

One way to speed up the removal and replacement of hatch covers was to join several boards together within rectangular steel frames to form a 'slab'. Captain A. Dunn of the New Zealand Shipping Co. devised and patented a simple slab hatch cover in 1924 and several of his company's vessels were fitted with them in that year and later. They were, of course, too heavy to be handled manually and therefore had to be lifted on and off with derricks and

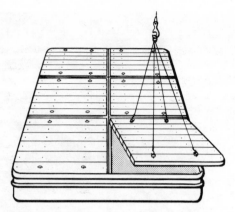

Fig. 1.5. Typical arrangement of slabs.

winches. Their use spread and even today a few ships equipped with them still remain in service (see Fig. 1.5).

It was not a universally popular arrangement, however. The fact that opening and closing depended upon the availability of deck power at short notice was seen as a serious disadvantage by some opponents. A further disadvantage was the difficulty of removing side hatches at sea for hold ventilation during fine weather, although this could be overcome by having a row of single boards along each side coaming.

In a typical hatch there were usually two or four slabs abreast across the width of the hatchway between each pair of 'king' beams (beams having a vertical web plate protruding through their upper horizontal flange to act as a board end stop). The next obvious development was therefore to join sets of slabs together to form a continuous cover, extending from one side of the hatchway to the other, which could be lifted out in one piece.

Very large slabs were awkward to handle, and were sometimes hinged so that they could be folded. It is interesting to note that, in Sweden, Captain E. von Tell held patents for folding wooden hatch covers as early as 1922 (see Fig. 1.6). Large slabs also required extensive stiffening and were thus very heavy. Once removed from a hatchway, they had to be put somewhere and sufficient deck area had to be available for this purpose, although this was less of a problem with folding slabs. Similar problems arose with hatch beams as well. The wider the hatchway, the longer and heavier the beams needed to span it and the more difficult, dangerous and time-consuming the process of shipping and unshipping them.

Until the 1930s, beams were invariably supported on bracketed seats attached to the side coamings; shipping them often involved much pushing and heaving because the derricks used to lift them could not readily be adjusted to plumb different parts of the hatchway. Once in place beams were secured

Fig. 1.6. Folding slabs being lifted.

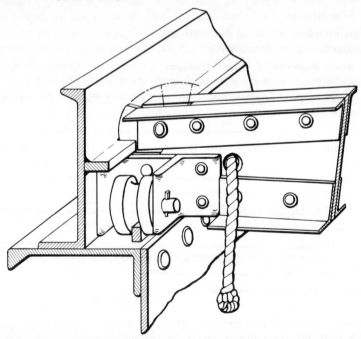

Fig. 1.7. One form of sliding hatch beam showing its arrangement at the hatch coaming.

with bolts which, in practice, were often left out – sometimes through negligence, sometimes because the beam seats which projected into the hatchway had become damaged – with the result that they were occasionally unshipped accidentally during the normal course of working cargo.

Sliding beams (Fig. 1.7), first introduced in 1929, were thus doubly beneficial – it was unnecessary to lift them out and almost impossible to unship them accidentally. These advantages were obtained by introducing a step into the side coamings, at the same level as the external horizontal stiffener, to provide a landing for the ends of the beams, each of which was fitted with a wheel or roller. It was thus possible to push them to the end of the hatchway to provide a clear opening. A horizontal bar was attached to each side coaming just above the top of the beam wheels to prevent accidental unshipping and a simple hinged key kept each beam in position when the hatch boards were being replaced.

Several patterns of sliding beams became available and were widely used. Together with slabs they were responsible for reducing both the time and labour required for opening and closing hatches. Significant savings were possible, but not enough to stop some shipowners from seeking a completely different alternative to the traditional wooden hatch cover. They were further encouraged in their search by considerations of maintenance.

1.3.3 Maintenance

The maintenance of wooden hatches was expensive in both labour and materials. Although the individual components were simple and inexpensive, there were so many of them in a typical hatch and they were so easily damaged through continual hard use that the total cost of replacements was often substantial. Wedges, which were vital to the security of a hatch, were individually very cheap but they were frequently lost or misused. Thus, thousands of replacement wedges might well have been supplied to a ship or made on board throughout its life.

Hatch boards too had to be replaced frequently. While the introduction of Cocks' reinforcing may have greatly reduced the damage to hatch boards, it did not eliminate it entirely. Moreover tarpaulins, upon which the security of traditional hatches depended above all, were so easily chafed and torn that in some ships one man could well have been employed continuously on their repair.

1.4 The coming of steel hatch covers

Nowadays, when large ships may have hatchways more than 25 m (80 ft) wide, the use of wooden hatch covers has died out almost completely. Leaving aside any considerations of safety and security, which are grounds enough to justify their disappearance, the physical labour involved in opening and closing very large wooden hatches rules them out as a practical option for modern vessels. Imagine lifting out beams 25 m long or replacing fifty or more boards across the width of every 1·5 m long section of every hatchway!

Perhaps more than anything else, the dramatic increase in ship size during the last decade or two has been responsible for bringing to an end more than two thousand years of continuous use of wooden hatch covers. While their shortcomings have long been recognized, and while alternatives have been available for much of this century, wooden hatches were still being installed in some ships at the end of the 1950s. But events since that time have lent considerable force to the arguments supporting steel hatch covers, notwithstanding that many of the arguments are the same as those first advanced more than half a century ago.

It is impossible to say who first suggested that hatch covers should be made of steel. Like many an idea that has eventually overturned established practices, the possibility that steel might be used for hatches was probably recognized long before any practical proposals based on it were advanced. Indications of an awareness of the potential of steel as a material for hatch covers were evident as early as the beginning of this century but there was nothing then to suggest that it would ultimately displace wood completely.

Fig. 1.8. USN collier *Cyclops*, built in 1910, discharging coal. The internal stiffening of the hinged steel hatch covers can be seen clearly.

Fig. 1.9. One of the ten Hogg-Carr steel hatch covers fitted in *Rose Castle* in 1915. Note their corrugated construction.

At that time, hatches were still fairly small and the steel covers that were envisaged appeared to offer no practical advantages over the wooden boards then in use. But some naval architects were prepared to advocate steel. For example, steel covers were specified in a novel cargo ship design which was published in 1908. Besides steel hatches, this vessel had divided holds and a double row of hatches which, it was claimed, would improve cargo handling – a claim which we should certainly recognize today.

Some of the earliest steel hatch covers were no more than small portable plates. They were used instead of wooden hatch boards and tarpaulins and wedges were still necessary. Ships were, however, occasionally fitted with hinged 'lids' large enough to cover an entire hatchway. One vessel so fitted was 12 500 tonne deadweight* (tdw) United States Navy collier *Cyclops* (Fig. 1.8) built in 1910. It was not until after the commencement of the First World War, however, that steel hatch covers began to gain any marked degree of acceptance.

One of the first ships to be fitted with Hogg–Carr Patent Steel Hinged Hatch Covers was the *Rose Castle*, a steam collier built in 1915 by Short

---

* 1 tonne = 1000 kg (metric ton); 1 ton = 1016 kg (long ton). The deadweight of a ship is its disposable load, i.e. full load displacement minus light displacement. It consists largely of cargo, but also includes fuel, stores, water and crew.

Bros. of Sunderland for the East Canada coal trade. For those times *Rose Castle* was a large vessel, approximately 140 m long, with a deadweight of about 11 000 tonnes, but it was her ten steel hatch covers that identified her as rather more than just another fairly ordinary cargo ship.

The Hogg–Carr hatch cover, in the form in which it was installed in *Rose Castle* (see Fig. 1.9), was constructed in one piece to seal the whole hatchway. It was mounted on hinges attached to the end coamings and there were neither portable beams nor tarpaulins. In these respects, it was similar to earlier hinged covers, but unlike them it was not stiffened by transverse bulb angles or channel bars. Instead, its main plating was dished so that it presented a deeply corrugated surface rather than a flat one and the undulations were sufficient to maintain its rigidity. Opening and closing was achieved by means of derrick and winch. In the open position the cover was held vertically against a stop provided for the purpose, and when closed the edge of the cover rested on a horizontal angle bar attached to the top edge of the coaming all round the hatchway. It was secured to the coamings by means of threaded steel toggles spaced about a metre apart and kept watertight by greasy hemp packing.

The principal advantage which Hogg–Carr covers offered over earlier steel hatches was their comparative lightness. Considerable weight savings were possible as a result of the absence of stiffeners. It was claimed that the covers installed in *Rose Castle* were each more than three tonnes lighter than would have been possible with any other steel type then available.

They were, however, heavier than equivalent wooden hatches, but in a typical ship of the time, the weight of a full complement of wooden boards and beams could have been around 25 tonnes and so the weight penalty borne by the slightly heavier steel covers was probably not as great as it may have appeared at first sight. Indeed, for very large hatchways, it was claimed that Hogg–Carr covers were lighter than orthodox wooden ones.

Hogg–Carr covers continued to be fitted in colliers and ore carriers for more than 20 years but their popularity which, although significant, had never been particularly widespread, had begun to wane well before the end of this period. They had numerous drawbacks. The most important of these was that they could only be opened or closed with a winch. Today this would not even be seen as a minor inconvenience but half a century or more ago, steam for deck auxiliaries was not always readily available. It was less of a problem in ore carriers than in general cargo ships because their cargoes were less likely to be damaged by inclement weather, thus making frequent opening and closing in port unnecessary. but it was an argument which persuaded many shipowners to keep to traditional hatch covers which, because they required nothing more than manual effort, could always be opened or closed. When closed, Hogg–Carr covers with their deep undulations were hardly an ideal platform for deck cargo even though they were strong enough.

Hogg–Carr covers were the first steel hatches to be manufactured in quantity to a standard design and therein lies their importance in the history of ship design. It was not until late in the 1920s that alternative types of any significance began to appear. This was when the eponymous name of MacGregor first became associated with steel hatch covers.

1.5 Steel hatches after 1927

The earliest satisfactory hatch cover produced by the brothers Joseph and Robert MacGregor, two naval architects from Tyneside, was of the hinged

Fig. 1.10. A hinged steel slab hatch cover. This was secured with tarpaulins, battens and wedges in the traditional manner, and was in use in the mid-1920s.

type. It consisted to two large steel slabs, each spanning half the hatchway and hinged at the side coamings (see Fig. 1.10). When opened by winch and derrick, the slabs rotated through 180° so that they rested on the bulwarks to form a convenient platform for working cargo (although obstructing the gangway along the deck). Like other slab hatches they were kept secure and watertight in the conventional manner and their hinges were cunningly built into the coamings so as not to impair the effectiveness of their tarpaulins. However, it was the announcement in 1928 of a horizontal rolling cover which did not require tarpaulins that marked the real beginning of the long association of the MacGregor name with steel hatch covers, which continues to this day.

The first horizontal rolling cover built to the design patented by MacGregor & King Ltd. was installed over the aftermost hatchway of the 9000 tdw motor ship *Sheaf Holme* which left the Sunderland yard of her builder, Wm. Pickersgill, in August 1929 amid rather more publicity than she would otherwise have merited. The event was important because it was the first time that a steel hatch cover had been put into service that was easily operated manually, yet neither depended on tarpaulins for watertightness nor did it forgo any of the strength with which previous steel covers had been endowed.

The earliest horizontal rolling covers produced by MacGregor and King (subsequently known as Macanking) came in two forms, although both were basically similar. The first was fitted in fairly small hatchways and consisted of two moveable sections fabricated out of steel plate stiffened by channel bars or bulb angles, with a skirt around their perimeters which rested on the coaming when the cover was closed. Each section was provided with four wheels which were attached to the skirt. They revolved about eccentric bushes which could be rotated so that the clear height of the lower edge of the cover above the coaming could be varied.

In the closed position, with the hatch secured for sea, the bushes were aligned so that the perimeter of the cover rested on the coaming throughout its entire length. To open the hatch, the bushes were rotated through half a turn with a marlin spike, thus raising the cover so that it could be easily pulled along on its wheels. This arrangement has become a feature of many MacGregor hatches since.

In the earliest covers the two sections, once raised onto their wheels, could be moved in a fore and aft direction with one pair of wheels rolling along a flat bar attached to the top of the coaming and the other pair rolling along a track extending from the ends of the coaming. In some cases the track was fixed to the deck and the wheels were attached to the end of the cover by 'legs'; in others, the track was constructed so that it was at the same level as the top of the coaming, although this second arrangement obstructed the deck to some extent. The purpose of both arrangements was to keep the cover horizontal as it moved. When stowage space at the ends of the hatchways was limited, the movable sections were designed to pivot at a critical stage in the opening process so that they could be stowed vertically and thereby take up less room (see Figs. 1.11 and 1.14).

In large hatchways, end rolling covers arranged in two sections were often impractical because there was rarely enough space at the ends of the hatchways to allow the sections to be stowed horizontally and the coamings were not high enough to permit vertical stowing. A second form of horizontal rolling cover was therefore offered in which the individual sections rolled transversely. A typical large hatchway, say 10 m long by 7 m wide, had at its midlength a heavy portable beam the purpose of which was to support the four sections of the cover. Each section was rectangular and occupied a quarter of the area of the hatchway with joints along the centre line and in the way of the transverse beam. The cover was opened by rolling each section outwards from the centre line on wheels and tracks similar to those used in fore and aft rolling covers (see Fig. 1.13).

When closed, both forms of cover were secured for sea by patent wedge-headed cleats which passed through slots in the horizontal stiffener attached to the top edge of the coaming, engaged lugs on the cover skirt and were locked in position with tapered pins. A seal was made between the bottom of the skirt and packing which rested on the coaming to keep water out; the cross-joints were also provided with very simple compression bar seals.

In service these hatch covers were reasonably watertight. They were easily operated by two men and they required little maintenance (there were no wedges to be lost or tarpaulins to be repaired). On the other hand their total weight was greater too.

While they attracted a lot of interest, it would be wrong to imagine that after 1929 every new ship, or even a majority of new ships, was fitted with them. However, as a result of arrangements with other firms in Europe and Japan, the number of vessels with Macanking steel hatches grew steadily.

Various refinements to the original designs which improved their efficiency or widened their scope were patented. Insulation fitted to the underside of each hatch section for use in refrigerated ships was one of the first modifications offered. This was a considerable improvement on the methods of insulating hatches then commonly in use, which required timber clad beams with insulated plugs to fill each beam space, in addition to the normal boards and tarpaulins. Opening and closing a traditional insulated hatch was thus even slower than an ordinary one, so that the advantages of combining the operations of beaming up, plugging up and battening down in the single act of pulling shut an insulated steel cover were clearly considerable.

Another early modification was introduced to make end rolling covers suitable for longer hatchways than could be conveniently served by two panel hatches. In order that more than two panels of reasonable size could be accommodated, the side coamings were castellated (or stepped) instead of horizontal. The forward and after sections of a three-panel cover were therefore secured at the same level to coamings of normal height, while the centre panel rested on coamings which were higher than normal, with its base aligned with the tops of the end sections. To open the hatch, the centre panel was first raised onto its wheels and then rolled forward or aft until it rested completely on one of the end panels, which was provided with the necessary track and securing points for this purpose. The remainder of the opening process was then similar to that for a two-panel cover, with the single section being rolled over one end of the hatchway and the tiered end and centre

Fig. 1.11. Two panel end rolling cover.

Fig. 1.12. Three panel end rolling cover with stepped coaming.

Early MacGregor (Macanking) hatch covers

Fig. 1.13. Four panel side rolling cover with a portable transverse beam.

Fig. 1.14. End rolling and pivoting cover.

sections being rolled over the other. A four-panel cover could be treated in the same way (see Fig. 1.12).

The ease with which early MacGregor hatch covers, even those installed over large openings, could be opened and closed was undoubtedly an important factor in their success. Numerous public tests were conducted during the early 1930s which consistently demonstrated the superiority in this respect of rolling steel hatch covers over any of the alternatives then available. For example, in 1931 it was shown that the four-panel end rolling covers fitted in the reconstructed passenger/cargo vessel *General von Steuben* owned by Norddeutscher Lloyd could be almost completely opened by two men within two minutes. The next year, a test was carried out in Japan aboard the *Nagoya Maru*, a 9000 tdw cargo ship with five hatchways fitted with side rolling covers. In this case, four men were found to be able to open the largest hatch, which was 10·1 m (33 ft) long and 6·1 m (20 ft) wide, in under three minutes, while with the assistance of a winch the same number of men recorded a time of less than two minutes. It was claimed that these times could be expected under normal operational conditions and were about a fifth of the time necessary to open a similar sized hatch of traditional pattern.

With present day experience of mechanical hatches these claims do not seem exaggerated. It must be remembered, however, that published opening times usually took no account of the time taken up in unfastening the cleats which had to be in place at sea. Opening hatches on arrival in port was begun by releasing the cleats and their replacement completed battening down on departure; both operations could be seen as directly comparable with the acts of knocking out or hammering home the many wedges securing the tarpaulins of an orthodox wooden hatch. It was therefore in the opening of hatches before a day's work, or closing them at the end of a shift or in rain, when full cleating was not required, that rolling steel hatches showed their handling advantages to best effect.

When the covers were closed in the course of normal cargo working, they were usually standing on their wheels. There was no danger in this so long as the ship was at a reasonable trim and roll-back preventer bolts were in place. In this position the covers were rainproof and the wheel clearance provided a means of additional ventilation to the hold. Moreover, at sea a hatch section could be temporarily released and raised onto its wheels in fine weather for hold ventilation thereby overcoming one of the objections to the use of steel hatches that had been advanced by their opponents.

The variety of ships in which MacGregor hatches were fitted during the 1930s was very wide. In addition to general cargo vessels, it included ore carriers and colliers, ferries, oil tankers – which in those days invariably had a dry cargo hold forward – and even ice-breakers. Moreover, many ships were equipped with steel hatches at the forward holds by owners who, while recognizing the improved safety of steel hatches, were unwilling to accept the weight penalties which they believed inherent in their use at every hatchway. One measure of the growing acceptability of steel hatches was the readiness of some shipowners to convert the hatchways of their existing vessels. The Irish Sea railway steamers *Duke of Lancaster* and *Duke of Rothesay* were amongst the first to provide evidence of this trend.

In 1938 the public debate on wood versus steel for hatch covers received added impetus from the loss of the steamer *Stancrest*. Harry Cocks and John Tutin, another steel hatch cover patentee, were among the main protagonists and it is interesting to note that the latter advocated a reduction in freeboard for ships with steel hatches – some thirty years before the principle was officially accepted.

There were no winners, however, save perhaps those owners who, during the Second World War a few years later, felt able to announce that their torpedoed ships had reached the safety of port only because they had 'floated home on their steel hatch covers'.

Besides the MacGregor Company, which was active by deed rather than by word, the 'pro-steel lobby', if thus it could be described, was supported by a growing band of advocates. Throughout the 1930s new designs for steel hatches were continually being published and patented. Some were successful, some not; some were supported by major organizations, some were little more than the enthusiasms of individuals. No survey of the development of hatch covers would be complete without a brief mention of some of the more important of these.

In the same year that the MacGregor rolling cover appeared, a German, T. Schwarz, also published a design for a similar rolling cover, His design seems not to have been popular however, and very little more was heard of the Schwarz hatch cover. The following year, Captain H. C. MacNeil of the Royal Australian Navy patented a steel cover of a type which had originally been installed in the seaplane carrier *Albatross*. His was a rolling cover constructed in sections, each of which was in the form of an inverted dish which provided adequate strength and rigidity without the need for stiffeners. The main advantage claimed for it was that the sections could be removed and nested. A wedge cleating arrangement was employed.

Also in 1929, Captain W. H. Sweney, a P & O dock superintendent, designed the steel hatch covers that were installed in the passenger liner *Viceroy of India*. They were hinged flat plates and were fitted only at Nos 3, 4, 5 and 6 hatches, primarily to avoid cluttering passenger decks in port with hatch boards, beams and tarpaulins. These covers had an interesting and novel cleating arrangement which consisted of an inverted 'G' clamp attached to the underside of the horizontal stiffener at the top of the coaming. With the cover in the closed position, the clamp was rotated until vertical and its screw tightened to press the edge of the cover against the coaming stiffener. Sweney hatches were not widely adopted although several other passenger vessels were fitted with them. The 'G' clamp cleats were, however, sometimes used in cargo vessels instead of bolts for securing deep tank covers.

In 1931, Captain E. von Tell, whose name has already been mentioned in connection with folding wooden slabs, produced a double-leaf folding steel cover. This consisted of two flat panels hinged together and at the end coaming. During opening and closing, the cover rotated about the coaming, rather like an inverted compound pendulum. Derricks were necessary for this operation, firstly to lift the panel remote from the coaming hinge and rotate it through 180° to rest on the nearer panel, and secondly to rotate the two panels together about the coaming hinge. Long hatchways were usually pro-

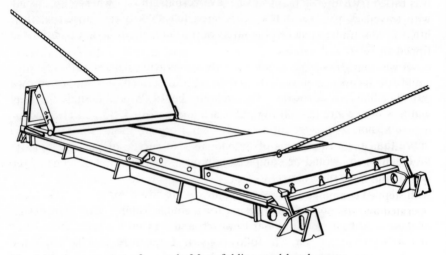

Fig. 1.15. Arrangement of an early Mege folding steel hatch cover.

vided with hinged panels at both ends. A number of British shipping companies, including Coast Lines and the General Steam Navigation Co., adopted von Tell covers – steel or wood – for a few ships.

In France in 1934, Captain Mege, of Louis Dreyfus Cie, patented folding steel hatch covers which were first fitted in the cargo ship *Louis L.D.* two years later. They consisted of double panels hinged together and to the end coamings, but, unlike the von Tell covers, they were opened by lifting the panel nearer the coaming hinge (see Fig. 1.15). As it was raised, the second panel was pulled towards the end of the hatchway, supported on two trailing wheels which ran along the coaming, until both panels were stowed together vertically. In this way a folding hatch cover could be opened in one operation instead of two. Closing was carried out by simply allowing the fold hinge to open under the combined weight of the two panels until both were in position resting flat on the coaming to which they were secured by bolts. In long hatchways, Mege covers were fitted to both end coamings. The folding principle that was incorporated into their design has become widely adopted, for both wire-operated and hydraulically operated folding covers. More than thirty ships were fitted with Mege covers before MacGregor & Co. became the sole licensee for their construction in 1949.

In Great Britain, at the same time that Mege's covers were being installed in the *Louis L.D.*, end rolling hatches designed by Tutin were being fitted in the 9000 tdw cargo ship *Starcross*. The Tutin hatch was made up of any number of sections resting on a horizontal coaming. To open it, each section was raised individually by means of a special jacking device that was fitted with traversing gear so that it could be moved along the coaming. Once lifted it was transferred on the traversing gear to the hatch end where it tipped to be stowed vertically. The principal advantage of the Tutin hatch cover was that quite large sections, which were too heavy to be pushed easily, could still be opened or closed without the use of winches. This was borne out the following year when Tutin covers with sections spanning a hatch width of 9·8 m (32 ft) and each weighing 2·5 tonnes were installed in the collier *Newton Moore*.

With the passing of time, the value of wholly manual operation decreased. Power for deck auxiliaries became more readily available at short notice and with it the need for Tutin covers declined. MacGregor's end rolling hatch with stepped coamings also gradually lost popularity to be replaced by wire-operated and rolling and pivoting covers with continuous coamings. Typical of these was the 12·2 (40 ft) long hatch installed in the cargo liner *Waipari* in 1941 by MacGregor & Co. Apart from chain links between each section, this displayed all the features which, a few years later, were to be incorporated in the well known single pull hatch cover, described in detail in Section 5.2.

In the years since the Second World War a wide variety of mechanical hatches have been introduced. One particularly interesting and novel type was invented in the 1950s by Mr Rene Caillet. This consisted of a series of close-linked transverse panels which, to open, rolled around a drum situated beyond the end of the hatchway. Pioneered by Ermans and subsequently developed by MacGregor, Caillet's design is the progenitor of the roll stowing cover now widely known as Rolltite (see Section 5.5). During this time, too, the use of mechanical hatches for tween-decks has also become increasingly common and Caillet was responsible for developing an early form of sliding tween-deck cover (Section 5.8).

Notwithstanding the enormous choice of hatch cover types – wire/chain/electric/hydraulic operated; end/side/rolling/folding/lift off; weather/tween-deck; flat/peaked top; single/double skin, etc. – now available to the modern shipowner, the single pull cover remains, for many people, the archetypal mechanical hatch and deserves to be described as the true heir to the traditional beam and boards.

CHAPTER TWO

Cargoes and Ships

Summary

In this chapter a description of the principal trades in which dry cargo ships are employed is given, together with a brief discussion of the properties of commodities and goods commonly transported by sea.

The main methods of cargo handling are outlined and the principal types of merchant ships now in service are introduced.

2.1 Introduction

A wide variety of goods enter international trade and many types of ship are used to transport them. Merchant ships are mobile warehouses whose many different forms have evolved as a result of attempts to balance on the one hand the need for suitable storage capacity, against on the other hand the need for mobility. Thus a ship constructed as a simple rectangular box of appropriate dimensions could provide an ideal space for storing containers, but it would be difficult to propel through the water, while an easily driven hull would offer relatively little usable cargo space. Ship design is largely a matter of resolving such conflicts to produce vessels that are suited to the services in which they will be employed.

The conditions in different trades vary so widely that it is impractical to build ships which can serve them all equally well. Instead, ships are built for specific purposes and although some are commonly described as 'general purpose', they each find employment in a fairly limited range of trades.

2.2 Dry bulk trades

Basic raw materials such as ores, coal and food-grains are usually transported in bulk and during the last three decades or so, the volume of such commodities entering international trade has expanded rapidly. The unit value of many bulk commodities is low, typically less than $100 per tonne, and trade in them is particularly sensitive to transport costs. Consequently ships to carry materials in bulk, efficiently and economically, have been specially developed.

Ships operating in the dry bulk trades must have good deadweight carrying capacity and they tend, therefore, to have full forms. Many of them are very large to take advantage of economies of scale and they usually have wide hatches and clear holds to facilitate loading and discharging. In most bulk trades the cost of time spent in port forms a significant proportion of the overall cost of transportation (see Section 3.1). Bulk carriers spend roughly half their time in ballast since most raw materials are exported from areas requiring little return cargo.

2.2.1 Iron ore

Iron ore is the most important of all the dry bulk commodities that are transported by sea; approximately 300 million tonnes are now shipped annually. It is exported mainly from South America, Australia, West Africa, Scandinavia and North America to Europe, USA, and Japan; trading distances average 5000 miles.

Like other natural raw materials, iron ore varies in composition according to its place of origin. In some cases it is transported simply as crushed rock, in others the basic ore is first transformed into enriched pellets. But in both forms it stows at 0·4–0·5 m³/tonne or 14–18 ft³/ton, and is thus one of the densest materials commonly carried in ships. This has important implications for the design of vessels for iron ore trades (see Section 7.1.4).

In recent years many new loading ports and discharging terminals have been constructed to meet the requirements of the growing iron ore trades and the development of new sources. Conveyors and chutes are used for loading, and rates of more than 6000 tonnes per hour are common. Grabs are usually employed at discharging terminals, and at typical installations they may each have a capacity of 20 tonnes and operate with a 40-second cycle, i.e. about 2000 tonnes per hour. There are physical limits to the rates at which ores may be handled by grabs, however, and during the last few years quantities of iron ore have been loaded and discharged as slurry (in semi-liquid form) through pipelines. Very high handling rates have been claimed for this method, although it is still not widely used.

Today, few vessels engaged in the carriage of iron ore have their own cargo handling equipment. Most are very large, usually in excess of 60 000 tdw and are employed on long term charter to the world's major steel manufacturers. In some cases they have been designed to their charterer's specifications.

2.2.2 Coal

The carriage of steam coal has provided employment for ships for centuries. Nowadays, however, practically all of the coal that is transported by sea is used in the manufacture of steel. Total shipments now exceed 100 million tonnes annually, mostly from Eastern Europe, North America and Australia to Western Europe and Japan.

Coal stows at 1·2–1·4 m^3/tonne (43–50 ft^3/ton). It is handled in a similar manner to iron ore, with conveyors and grabs, and is carried mainly in bulk carriers of more than 40 000 tdw.

2.2.3 Grain

The term 'grain' includes wheat, maize, oats, barley, rye and a variety of other seeds of which more than 100 million tonnes are transported annually by sea. Both the demand for grain and the available supply vary from year to year as a result of largely unpredictable factors such as the effect of weather conditions on local crops. The principal grain exporting areas are the USA and Canada (including the Great Lakes), Australia and South America. Most other regions of the world are, always or occasionally, importers.

Grain stows at 1·2–2·0 m^3/tonne (43–72 ft^3/ton). It has a low angle of repose, readily settles in a ship's hold and shifts if it is not properly stowed. Consequently its carriage is governed by internationally agreed regulations, the main provisions of which are concerned with the stability of the carrying vessel and the arrangement of its holds. Grain in transit can be easily damaged. It is particularly important that sea water does not leak through the hatches or condensation form in the holds.

Traditionally grain has always provided employment for 'tramps', i.e. ships which accept whatever suitable employment offers. Today many small and medium size bulk carriers and multi-deckers (traditional general cargo vessels) are employed in its carriage, mainly on voyage charters or short time charters. It is a trade which favours relatively small ships because of the wide variety of ports, many with limited facilities, which are used and because of its seasonal nature. Grain is invariably loaded by conveyors and chutes and at most ports it is discharged pneumatically. In some parts of the world, however, grabs (sometimes attached to the ship's own derricks or cranes) or even manual labour may be used.

2.2.4 Bauxite and phosphates

Roughly 50 million tonnes each of bauxite and phosphates are transported annually by sea. Bauxite, which is the raw material for the manufacture of aluminium, is exported mainly from the West Indies and Australia to the USA, Japan and Europe, while phosphate rock, which is used as a fertilizer, is exported principally from Morocco and the USA to Europe, Japan and Australasia. Both stow at 0·7–1·1 m^3/tonne (25–40 ft^3/ton) and are usually carried in medium size bulk carriers or multi-deckers.

2.2.5 Forest products

'Forest products' is a general term used to describe logs, sawn lumber and manufactured products such as wood pulp, paper and hardboard. World-wide shipments amount to more than 100 million tonnes annually. Most of the timber transported by sea is softwood, used in the construction in-dustries and for the manufacture of paper. The principal exporting areas are North America, Scandinavia and the USSR, supplying Europe and Japan.

Traditionally, timber has been transported in small cargo vessels and coasters. In recent years, however, the practice of 'banding' sawn lengths of timber together to form easily handled, regular 'packages' has led in-creasingly to the employment of bulk carriers in the timber trades. Bulk carriers of 50 000 tdw now carry packaged timber and ships of this size are being built specifically for this purpose (see Section 7.1.6). Stowage factors vary widely, from 1.5 m³/tonne (54 ft³/ton) for packaged lumber to 3·0 m³/tonne (110 ft³/ton) for logs.

The carriage of timber is governed by international regulations which permit a reduction in a vessel's normal freeboard when timber is on board, provided that it is properly stowed and secured both in the holds and on deck, where it provides additional buoyancy and protection for the hatch covers.

2.2.6 Steel products

Some 60 million tonnes of steel plates, coils, bars and other products are exported annually, principally from Japan and Western Europe, and are commonly carried in bulk carriers, general cargo and Ro-Ro vessels. With stowage factors of 0·3–0·6 m³/tonne (10–20 ft³/ton) they occupy little space and can therefore be carried in the large, deep holds of bulk carriers with none of the stowage problems which would accompany the carriage of bulkier goods. Many of these products are used in the manufacture of consumer durables and their value is considerably reduced if they are damaged in transit. Rust is a common form of such damage and it is there-fore essential that the hatches of the carrying vessel do not leak and that condensation is not allowed to form in the holds.

2.2.7 Other bulk commodities

Numerous commodities in addition to those mentioned above are trans-ported in bulk by sea, in bulk carriers or multi-deck general cargo vessels. Among the most important of these are: manganese ore, non-ferrous ores, iron pyrites, scrap iron, salt, sulphur, gypsum, mineral sands and petroleum coke. Together these amount to over 100 million tonnes transported annu-ally worldwide.

2.3 Bulk liquid trades

Bulk liquids have been transported by sea for nearly a century. Today, oil (both crude and refined products) is the most important single commodity carried in ships, exceeding in volume the combined total of all other commodities entering international trade. Crude oil tankers are now the largest ships afloat with deadweight capacities of up to 550 000 tonnes.

During the last few years ships have been developed to carry liquified gases in bulk and others now carry liquid chemicals. These incorporate many special features which are imposed by the (often dangerous) properties of the cargoes which they carry. Detailed description of bulk liquid trades and the vessels employed in them is beyond the scope of this book, however.

2.4 Break-bulk general cargo

Manufactured, semi-finished and many other general goods commonly constitute liner cargoes which have traditionally been transported in 'break-bulk' form. Each item is individually packaged in a carton, crate or bale and must be handled separately as it is loaded into the ship, stowed in its hold and finally discharged at its destination. The unit value of general cargo is usually much higher than that of bulk cargoes and $500 to $5000 per tonne is typical. Thus even though the total annual general cargo carried in ships amounts to only about 300 million tonnes, it is easily the most valuable sector of seaborne trade.

Scheduled liner services have been offered for a little more than one hundred years and are now operated over many routes throughout the world. Individual consignments of cargo are small, usually considerably less than a shipload. Thus on any voyage a cargo liner might carry a wide range of goods for many different shippers. Moreover the conditions that exist in different trades are extremely diverse and no single trade can be regarded as typical of liner services in general. Some services are operated between industrial countries and a large proportion of the cargo carried in each direction consists of manufactured goods. Other routes link industrial and developing countries where the flow of manufactured goods in one direction is not matched by a similar return flow. Liner cargoes loaded in developing countries often consist of agricultural produce and relatively high value raw materials shipped in small quantities.

Break-bulk general cargo ships are small by the standards set by other types of ship today and do not often carry more than 15 000 tonnes of cargo (although note Section 3.3). There are few significant economies of scale. Typical general cargo liners spend roughly half their time loading and discharging in port so larger vessels, employing traditional cargo handling methods, would spend an even greater proportion of their working lives in this way. Break-bulk general cargo handling rates depend very much on the nature and variety of the goods which are being worked – 10–20 tonnes per gang hour is typical. A gang working in one hold may consist of from 10 to 20 men depending on the arrangements at particular ports.

General cargo ships must be capable of receiving a wide range of goods with different properties and destined for different ports. Some goods are fragile and must be handled carefully, some are noxious or odorous and likely to taint others already on board, some are especially bulky, others must be kept cool. It is not possible to give a complete list of all the goods carried in general cargo ships but the following selection indicates their variety:

Table 2.1. Typical general cargoes

Goods	Typical packaging	Important carrying considerations
Plant, machinery, and manufactured goods	Uncased/crates	Heavy, bulky and easily damaged
Chemicals	Bags/drums	Noxious properties, risk of spillage; often carried on deck
Processed foods	Cartons	Easily damaged, must be kept dry; often pilfered
Liquor	Cartons	Often pilfered
Fruit	Cartons	Requires refrigeration; persistent odour
Tea	Chests	Readily absorbs taint
Cotton	Bales	Liable to heat and ignite spontaneously
Hides	Bales	Odorous and vermin infested
Copper	Ingots	High value
Oilseed cake	Bags	Liable to heat up

Stowage factors vary widely, from 0·3 m³/tonne (11 ft³/ton) for metal ingots to 6 m³/tonne (210 ft³/ton) for bulky or awkward packages and uncrated goods. A typical general cargo as loaded in a ship might stow at an overall rate of 2·0–2·5 m³/tonne (70–90 ft³/ton).

Break-bulk general cargo vessels are equipped with their own cargo handling gear – derricks or cranes – but handling is highly labour intensive and expensive in high wage countries. Since the 1960s general cargo liners have been replaced by unit-load carriers on many routes.

Liner freight rates are regulated by cartels generally called 'Conferences' and, because of the scheduled services and the nature of the cargoes, are invariably higher than those in the bulk trades which are largely determined by market forces. Most conferences adopt the policy of charging 'what the traffic will bear' which results in a complex scale of charges covering all of the many different classes of goods that may be transported over a particular route.

2.5 Unitized cargo

Many of the handling problems associated with the carriage of break-bulk cargo have disappeared with the introduction of unitization. Goods are assembled into some form of standard unit, perhaps at an exporter's works, and the whole is transported, whether by land or sea, to its destination without further disturbance to the goods contained within it. In this way the number and variety of items to be handled is reduced and high handling rates can be achieved with mechanized equipment. Moreover, by this means international transportation may be regarded as an integrated system, each element of which can be designed to achieve minimum overall transport cost.

Ships specially constructed for the carriage of unit loads are now common. Many of them are much larger than the break-bulk vessels that they have replaced because higher handling rates have given rise to significant economies of scale.

2.5.1 Containers

Containers are the most important of the units that have become common over the last two decades. They were first used more than 60 years ago but it was not until the late 1950s that their full potential began to be realized with the introduction of standard sizes, and there are now few major liner routes on which containers are not employed.

Standard ISO (International Standards Organisation) containers are available in a number of complementary sizes, the most common of which has the nominal dimensions $6·06 \times 2·44 \times 2·44$ m ($20 \times 8 \times 8$ ft). Many containers now in regular use are $2·59$ m high ($8·5$ ft) (see Table 9.3). They are used for the carriage of any goods that can be fitted inside them and a variety of special containers, conforming in overall dimensions to the standard sizes, are also available for the carriage of liquids and refrigerated goods.

Some container services, like the break-bulk services that they have replaced, are organized so that ships call at several loading and discharging ports. But in others, overseas container shipments are routed through one main port at each end and outlying ports are served by subsidiary feeder services (or overland). This has led to the development of two types of container ship: those which may be described as mainline vessels which are usually large and fast, and feeder ships which are smaller and slower.

2.5.2 Pallets

Pallets are small rectangular platforms, usually made of wood, to which items of general cargo can be secured to form a unit that is easily handled by fork lift truck. Their dimensions are not completely standardized although most pallets are around $1\cdot2 \times 1\cdot0$ m (48×40 in). Operators on a few routes have adopted pallets in preference to containers to provide fully-integrated services. They are also often used in break-bulk trades to facilitate cargo handling.

Palletized cargo can be carried in any general cargo vessel provided that its decks and tween-deck hatches are strong enough to support fork lift trucks. Several specialized pallet carriers have been built, equipped with side doors to allow horizontal loading.

2.5.3 Roll-on/roll-off trades

Roll-on/roll-off (Ro-Ro) vessels are employed on certain general cargo services. Until the late 1960s these were predominantly short sea routes, often combined passenger/cargo services (see Section 8.3.1), but Ro-Ro vessels are now being increasingly used on transoceanic routes including, for example, Europe–North America and Europe–Australia.

The Ro-Ro configuration has been developed primarily to facilitate rapid loading and discharging and it is not itself a means of unitizing cargo. However, it can be adapted to meet the requirements of a variety of standard units, including containers which may be carried on trailers or by fork lift trucks, pallets, vehicles, loaded lorries as well as uncrated export cars, and large indivisible loads such as heavy plant.

2.5.4 Barges

A method of unitizing cargo which has been introduced on a few trade routes employs standard rectangular barges (lighters) of several hundred tonnes capacity, each of which is filled with break-bulk cargo in the traditional way, and is then loaded aboard a specially constructed barge-carrying vessel by crane or elevator, for transportation to its port of discharge (see Section 8.4.2). This arrangement allows the expensive barge carrier to be turned round rapidly in port but the barges themselves can usually only be loaded or discharged at the normal rates for general cargo. Barges are claimed to be suited to services linking developed and developing countries, particularly those having extensive inland waterway systems.

2.6 Cargo handling

The rate at which cargo is loaded aboard or discharged from a ship has a significant bearing upon the overall cost of its transportation from one place to another. Excessive time in port deprives consignees of the use of their goods, and ship operators of the use of their vessels. It is not surprising, therefore, that throughout modern times, the improvement of cargo handling methods has been a constant aim of many of those concerned in the operation of ships. Important advances, some of which are outlined below, have been made during this century, particularly in recent years.

Every cargo handling/transfer system consists of a number of readily identifiable elements. There is a 'source' and a 'sink' between which material is transferred; these may be a quayside storage area and a ship's hold. Then there is the material itself, which may take any of the forms described in the

Fig. 2.1. Modern twin hatch general cargo vessel equipped with cranes. Note the Stülcken derrick between Nos. 2 and 3 hatches. A general arrangement of this vessel is shown in Fig. 8.1. The ship is at a typical cargo liner berth, with dockside cranes and transit sheds. Cargo can also be handled overside into barges.

foregoing sections of this chapter. Finally there is the medium by which the material is transferred, which, in practice, may be manual labour, specially-designed equipment or some combination of the two. In an efficient system, all of these elements must be properly matched.

The earliest efforts to increase handling rates were concentrated mainly on the last-mentioned element, the transfer medium. They led to the development of a wide range of mechanical equipment, cranes, conveyors, etc., which have substantially improved loading and discharging rates, especially for homogeneous (usually bulk) cargoes. General cargo handling, however, has not benefited to such a great extent from such developments. In liner trades, the principal impediment to high handling rates has always been the large variety of packagings used for general cargo, so that significant improvements have only become possible by reducing the number of different forms in which goods are presented for shipment. Thus it is only with the adoption of unitization that general cargo carriers have achieved high transfer rates and been able to take advantage of handling techniques similar to those which have been developed for homogeneous cargoes.

2.6.1 Derricks

Derricks have evolved from the spars of sailing vessels as a means of lifting and transferring cargo between quay and hold. They can be rigged in a variety of ways but the most common arrangement is the 'union purchase' in which the falls (lifting wires) of two adjacent derricks are led from their respective winches through blocks at the derrick heads and are attached to the same cargo hook. One of the derricks is topped (raised to its required height) to plumb the hatchway while the other is topped and slewed to plumb overside. Both are secured in position by means of wire guys and during normal operation remain fixed. Cargo to be loaded is assembled and slung on the quayside immediately below the overside derrick, by which it is raised until it is above the ship's rail. From this position the weight is gradually transferred to the inboard fall, traversing the deck until it can be lowered directly into the hold. The process is reversed for discharging.

Union purchase provides a simple and speedy means of loading and discharging both general cargo in rope slings or on pallets, and bulk materials in canvas slings, provided that each lift is fairly light. A pair of derricks, each having safe working loads of 5 tonnes when rigged singly, would be able to lift no more than $1\frac{1}{2}$ tonnes in union purchase because of the forces which this arrangement sets up in the standing guys. In the past this limitation has not been a serious disadvantage but nowadays it effectively rules out union purchase for handling unitized cargo such as containers or packaged timber. Another important drawback is the difficulty of controlling the trajectory of the load. With unskilled winch drivers slings of cargo are often damaged by contact with hatch coamings and other parts of the ship's structure which may also suffer damage themselves. Moreover, it is largely impracticable to 'spot' (land) slings or cargo in any required position since the positions of the derricks are fixed.

Swinging derricks can handle heavier lifts with greater precision than is possible by union purchase. In one of its simplest and oldest forms, a single derrick is topped to a convenient angle and the free ends of its guy tackles are led to the whipping drums of a winch. The derrick can then be slewed but not raised or lowered (luffed). A number of different types of swinging derricks are now available and in common use. Most offer substantial advantages over the traditional arrangement, including the capacity to luff and slew simultaneously in order to facilitate spotting. In most cases this is achieved by substituting two widely spaced, powered topping lifts for the manœuvring guys. Safe working loads of 20–25 tonnes are typical for modern swinging derricks, so that they may lift 20 ft containers. They are frequently installed in cargo liners and small- and medium-sized bulk carriers.

Heavy-lift derricks were invariably part of the equipment of traditional general cargo vessels, even though they were not often used in service. Usually they were no more than large swinging derricks, of orthodox pattern, having a capacity of perhaps 75–100 tonnes. Their operation, by means of guys, topping lift and fall led to separate winches, was slow and cumbersome and often disrupted cargo work throughout the ship. But they had the advantage of cheapness and in trades offering few heavy loads their drawbacks were acceptable. Many are still in service.

During recent years there has been a great increase in the number and size of heavy indivisible loads being transported by sea. This has led to the development of improved handling equipment and ships designed specifically for their carriage. One of the first of the improved types of heavy derricks which are now available, and one of the most successful, is known as the Stülcken derrick (see Fig. 2.1). This is a heavy tubular derrick which, in its stowed position, is supported vertically between a pair of heavy samson posts arranged athwartships and canted outboard. Twin topping lifts, one to each samson post, provide slewing and luffing control and twin falls, one on the forward side of the derrick and the other on the after side, allow the derrick to be used as required, either at the hatch immediately forward of the samson posts or at the one immediately abaft them. Stülcken derricks capable of lifting more than 500 tonnes have been installed in ships. A number of vessels, most less than 7000 tdw, designed specifically for the carriage of heavy loads up to 1000 tonnes, have entered service in recent years (see Section 8.4.3).

The heaviest loads currently transported by sea are modular sections of offshore oil production platforms, some of which weigh more than 2000 tonnes. These are usually jacked up onto trailers and rolled aboard flat

Fig. 2.2. Hallen universal swinging derrick mounted on a bipod mast. The twin topping lifts, attached to the derrick head just off the bottom of the photograph, are led to independent winches on the mast house top.

pontoons for transportation offshore where they are erected by means of purpose-built crane barges and sheerlegs.

2.6.2 Cranes

Cranes were first installed in ships around the beginning of the present century. They were small, often hydraulically powered, and were used principally for handling passenger baggage and stores. Later, cranes were used in warships for handling seaplanes. It was not until the late 1940s, with the construction of ships such as the Swedish Johnson liners *Lions Gate* and *Golden Gate* that cranes were extensively fitted in general cargo vessels. Even so, derricks continued to be standard equipment in most merchant ships.

Early cranes were expensive compared with derricks; they were slower to operate and required more highly skilled drivers. Moreover their capacity and outreach was limited and it was often necessary for them to be fitted on transverse rails so that they could be manœuvred outboard of the hatch coaming in order to plumb both overside and the hatchway. For many ship operators, their ability to spot loads was insufficient advantage to outweigh these drawbacks.

With the increase in the carriage of unitized cargo, the practical value of the crane's accurate spotting ability has become more apparent. Thus cranes, often having a capacity of 25 tonnes and sufficient outreach to plumb two adjacent hatchways as well as overside, are now commonly found in ships. But where cost and other considerations are not overriding, owners of vessels frequently calling at ports which are well equipped with quayside cranes where labour is known to be unskilled may still prefer derricks to cranes, although today this argument is probably losing its force.

Crane or derrick falls (lifting wires) may be used for opening hatch covers, notably the direct-pull covers described in Section 5.4.

2.6.3 Other shipboard cargo transfer equipment

Shipboard alternatives to derricks (and cranes) have been available for special purposes for many years. One of the earliest, developed by Wm. Doxford Ltd, was installed in a number of small colliers (including the *Hermann Saubier*) before the First World War. It consisted of several inclined conveyor belts linking the holds with a gantry supporting overside chutes at the after end of the ship. A similar arrangement has been employed for more than half a century in Great Lakes vessels and in a few deep sea vessels too. Here, bulk material is allowed to flow, through inner bottom doors, from each hold onto a conveyor running through a duct in the double bottom, by which means the material is transported to one end of the vessel. There it is transferred to a vertical conveyor. This raises the cargo above the weather deck where overside chutes transfer it ashore.

Conveyor systems have also been installed recently in a number of wood-chip carriers. Because of the low density of their cargoes these vessels have broad and deep hulls and in many cases it has been found impracticable to build adequate quayside discharging equipment at the berths at which they normally call. Thus they are equipped with travelling cranes which are used to grab out the cargo from the holds and transfer it to conveyors running along the weather deck. These transport the chips to a fixed gantry supporting overside chutes.

Ships equipped with conveyor or other special cargo transfer systems are sometimes called 'self dischargers' although this description could easily be applied to any vessel having its own cargo gear.

Fig. 2.3. Cranes secured for sea in the modern 28 000 twd bulk carrier *Wayfarer*. Note the single pull, pan-type hatch covers, and the crane grabs stowed on deck.

2.6.4 Gearless vessels

Although the majority of dry cargo vessels have always been equipped with their own cargo handling gear in the past, today many ships rely entirely on shore facilities. Whether or not shipboard cargo gear is necessary depends very much upon the trades in which a particular vessel is likely to be employed, the port facilities and the attitudes of individual operators.

The largest bulk and ore carriers are often employed on long-term charter and, because of their size, can only call at a relatively small number of loading and discharging terminals where extensive handling facilities exist. Thus there is no need for them to have their own cargo-handling equipment. Small- and medium-sized bulk carriers, however, may be employed in a variety of trades requiring them to call at a wide range of ports, some with adequate facilities, some without. The availability of derricks or cranes increases the flexibility of such ships, allowing them, perhaps, to accept charters which otherwise they would have to refuse. This flexibility is achieved only at the cost of having handling equipment idle for much of its life and it is a matter for the operator's judgement to determine an acceptable level for such cost.

General cargo vessels have invariably been well endowed with cargo-handling equipment in the past but many of the container ships that have replaced them have none at all. Purpose-built container berths have been established in most important ports of the world and special cranes and other facilities with which they are equipped have rendered shipboard container handling gear largely unnecessary, although sometimes it may be fitted temporarily while port facilities are under construction. Moreover, because of the speed and precision with which heavy containers must be transferred between ship and shore, container cranes are complicated pieces of equipment and hence initially expensive. Thus, in many services, the low utilization of shipboard installations is normally unacceptable.

2.6.5 Temporary cargo gear

Gearless ships, calling at ports with inadequate quayside facilities, may sometimes make use of a variety of temporary cargo-handling equipment including small mobile cranes and portable pneumatic grain elevators. The use of such equipment can never be regarded as more than an occasional expedient, however.

2.6.6 Horizontal loading

Train ferries were amongst the earliest Ro-Ro vessels and many have been built since the middle of the 19th century. The simplest types are little more than self-propelled pontoons with sets of rails running along the weather deck; in other types, the train deck is enclosed by a superstructure with

Fig. 2.4. A typical container and Ro-Ro berth at Southampton. Note the extensive storage area for containers and export cars, and the straddle carriers near the crane. The vessel alongside is *Atlantic Causeway*. The bridge ramp (or linkspan) can be adjusted for the tidal height and the level of the stern door threshold.

passenger accommodation. With one or two notable exceptions employing elevators, rolling stock is carried on one deck only. End loading and discharging at specially constructed berths is usual and some vessels can accept rolling stock over both bow and stern.

Many of the Ro-Ro vehicle ferries and dry cargo vessels that have entered service in recent years can also load and discharge over both ends. This facility is particularly useful in ferries because it allows vehicles to be driven on board forwards and driven off again in the same direction, thus obviating the need for turning space or a turntable within the vehicle deck. This is a less important consideration in Ro-Ro cargo vessels, owing to the small number of accompanied cars that they carry.

At the present time, most end loading Ro-Ro cargo vessels have axial stern ramps (i.e. on the ship's centre-line) wide enough for two-way traffic and designed to link up with adjustable bridge ramps (linkspans) at specially-constructed berths (see Fig. 2.4). Some of the more recent are fitted with angled ramps, usually on the starboard quarter, which allow them to lie alongside unmodified general cargo berths, or slewing ramps which can be used on either quarter or axially (see Fig. 6.20). Cargo may be transferred on board from quayside storage areas by means of a variety of vehicles including self-propelled fork lift trucks, straddle carriers which are large vehicles that carry their load underneath the bridging between their wheels (see Fig. 6.4) and trailers, towed by tractors, some of which remain on board with their load

Fig. 2.5. Palletized cargo being loaded by fork lift truck through the side door of a small pallet carrier. See also Fig. 6.46 and 10.6.

(see Section 6.1). A number of Ro-Ro vessels are equipped with their own handling vehicles but this is not often necessary on routes over which Ro-Ro cargoes are regularly transported.

Ships designed for the carriage of export cars, pallet carriers and other side loading vessels often have several access doors along their length. In some, a ramp at each entrance allows cargo to be transferred directly into its stowage position by fork lift truck or similar vehicle (or driven aboard in the case of export cars). Other vessels have only a small landing area at each doorway where cargo can be transferred between fork lift trucks operating on the quay and those operating on board as illustrated in Fig. 2.5. Internal ramps and/or elevators provide access to different decks.

2.6.7 Port handling facilities

In recent years the advances which have been made in ship design and transportation technology have led to radical changes in the facilities offered by ports and have rendered a large number of traditionally arranged and equipped berths obsolete. The great lengths of quay space, with lines of cranes and transit sheds, which were the chief characteristics of every major general cargo port until the 1960s, have now been largely superseded by unit-load berths whose most outstanding feature is an extensive parking and marshalling area, mainly for containers.

The loading and discharging of break-bulk general cargo is not materially affected by the existence or absence of adequate quayside facilities. Shore cranes may be more convenient to use than ship's derricks but they do not speed up the transfer process to any significant extent since so much time is spent in making up, breaking down and sorting slings of cargo in transit sheds and holds, and securing for passage. There is almost no limit to the variety of packages and goods that can be handled at a traditional general cargo berth but this degree of flexibility is achieved only at the expense of handling rates which rarely exceed 500 tonnes per day. In contrast, the high handling rates that are possible at container berths, where only standard-sized container units can be accepted, are achieved at the expense of flexibility.

Containers awaiting trans–shipment are stored until required in parking areas adjacent to the berth, where they are usually arranged in accordance with some predetermined plan so that they may be loaded in the correct order – allowing for weight, contents and port of destination. They are moved between quayside and parking area by straddle carriers or other purpose-built vehicles or trailers, and transferred to the ship by large container cranes. A typical berth might have two such cranes, each with a capacity of thirty or more loaded containers per hour, corresponding to a daily rate of some 10–15 000 tonnes.

Specialized facilities for handling bulk materials have long been available

Fig. 2.6. The deck of the 1550 TEU container ship *Dart Europe* during loading operations. Note the cell guides visible through the open hatchways and the container stacking points on the pontoon hatch covers. Pontoon covers are visible stacked on the quay.

at many ports, but during the last two decades advances, comparable to those which have so changed the handling of general cargo, have also been made in the field of bulk cargo handling. Many of these advances have been precipitated by the increase in the volume of bulk materials which are transported by sea and the increase in the size of the ships used to carry them.

In many parts of the world, new loading and discharging berths and terminals, even completely new ports, have been constructed to promote particular trades. Many of them are capable of handling large vessels, although only a fairly small proportion can accommodate the very largest that are now in service with drafts of more than 15 m (50 ft). Unlike general cargo and container berths, bulk terminals are usually employed either for

Fig. 2.7. The 150 000 tdw OBO carrier *Muirfield* discharging iron ore at Redcar. This vessel is equipped with large side-rolling hatch covers, as is typical on such vessels. The unloader can discharge 2000 tonnes per hour to the conveyor system on the quay, which can deliver to rail wagons or stockpile. The derricks are for handling oil hoses; the cargo oil pipes can be seen outboard of the coamings.

loading or for discharging but not usually for both, such is the nature of bulk trades. Chutes and conveyors, fed by railway wagons or linked directly to adjacent stockyards, provide the most common means of loading bulk commodities, arrangements which are, in principle, often similar to those which have been employed at coal staithes for over a century. Relatively simple facilities can offer high handling rates, 50 000 tonnes per day or more.

Discharging terminals are often sited close to the plant where the bulk materials will be processed e.g. steel works, sugar refineries, or flour mills, and high handling rates are again necessary if large vessels are to be turned round in a reasonable time. Grain is usually discharged pneumatically, but grabs of various kinds are more commonly employed for other bulk materials. The largest grab installations are the unloaders used at iron ore berths, where the total weight of a grab and its contents may exceed 40 tonnes and the hourly capacity may exceed 2000 tonnes (see Fig. 2.7). In general appearance they are not unlike container cranes and some are operated with the assistance of a computer. The hopper-shaped holds of most bulk cargo ships assist grab discharging (see Fig. 7.1), although bulldozers are often necessary to clear the tank top (the inner bottom).

Bucket wheel reclaimers and bucket chain unloaders are sometimes used to discharge bulk materials where their higher continuous speed outweighs their higher cost.

But for all the development that has taken place in recent years, the standard of facilities, both equipment and labour, offered by different ports around the world varies enormously. At some, conditions may still be primitive, while at others congestion may be a perennial problem, which renders theoretical estimates of potential cargo handling rates of shipboard or other equipment of little more than academic interest.

2.6.8 Cargo stowage

Stowing a break-bulk general cargo is a task which demands considerable ingenuity and much manual skill. It must take account of the diverse properties of the many different goods making up a complete cargo, their possible interactions, their loading and discharging order, and once a stowage arrangement has been proposed, each item must be manœuvred into its designated position within the ship. In traditional general cargo vessels this entails a great deal of manhandling, especially of cargo which has to be stowed away from the hatch square, that is only partly alleviated by the assistance of fork lift trucks or other mechanical aids (see Fig. 2.8).

A principle of general cargo stowage which has been closely observed in the past is that the amount of unused (or unusable space) within a hold should be kept to a minimum. Dunnage (rough timber) is used to facilitate stowage and herein lies much of the skill of the stevedore. But the modern practice is to sacrifice space, where by so doing higher handling rates are achieved. Thus, in cellular container ships only space within the hatch square is used for cargo and in Ro-Ro vessels a great deal of space may be 'lost', especially when goods are not removed from their trailer chassis.

The stowage of bulk commodities (other than packaged timber which requires lashings) creates few problems. Some medium-density materials such as coal and grain must be trimmed, i.e. the upper corners of the holds under the deck must be filled, and machines for this purpose, which fit onto the loading chutes, are usually available at loading terminals. In many bulk carriers the holds are so shaped that trimming is unnecessary.

Fig. 2.8. Break-bulk cargo stowed in the tween-deck of a modern twin hatch ship, the same vessel as shown in Fig. 2.1.

2.6.8.1 *Bale and grain capacities*

The total volume of cargo spaces is usually given in two forms. Grain capacity is their total volume measured to the shell plating, including all space between the frames and beams; bale capacity is their total volume inboard of the frames (or sparring where fitted) and beams, typically about 10% less.

2.6.8.2 *Stowage factor*

Stowage factor is usually given as the volume, in m³ or ft³, occupied by one tonne of a particular good or commodity. In practice it includes an allowance for 'broken stowage', i.e. that proportion of the space available for cargo which is wasted. For homogenous bulk commodities, broken stowage is usually small, but for irregular packages, such as might be found in a typical break-bulk cargo, it may be substantial, perhaps as much as 40–50% in some cases. Broken stowage tends to be high in Ro-Ro vessels, especially where cargo remains on trailers.

2.6.8.3 *Measurement tonnage*
In liner trades, freight is usually charged on 'weight tons' carried for heavy goods and on 'measurement tons' for light goods. In some trades a measurement ton is 40 ft^3 although in others the m^3 (35·32 ft^3) is more usual.

2.7 General arrangement of ships

2.7.1 Hull form and dimensions
Hull form design has been one of the principal preoccupations of naval architects for many years and procedures have been devised by which the 'best' hull that meets a given specification can be selected. For merchant ships, the selection criterion is almost always an economic one.

The overall dimensions, proportions and shape of a ship's hull depend on a variety of hydrodynamic and service considerations. For example, its maximum dimensions may be constrained by physical limitations such as water depths and lock widths at ports at which it will call. One common type of bulk carrier today, known as the 'Panamax', is designed to have the greatest possible deadweight-carrying capacity consistent with economic operation, within the dimensional constraints set by the locks in the Panama Canal. Breadth is the usual limitation and 32·2 m (106 ft) is the maximum permissible. Such vessels have very full hull forms. But the fuller they are (i.e. the more nearly the hull approximates a rectangular box shape) the greater their resistance to motion through the water, the greater the power required to propel them at a given speed, and hence the greater the corresponding fuel consumption.

Container and general cargo ships usually operate at higher speeds than bulk cargo vessels, owing to the higher value of their cargoes. Thus they have finer lines and, for any given set of overall dimensions, they offer less internal cargo space and less deadweight capacity. To overcome the lack of usable internal cargo space, container ships invariably carry a significant proportion of their cargo on deck, while in many Ro-Ro vessels and vehicle carriers where similar problems arise, the main cargo spaces are located above the load waterline. The extent to which this solution is feasible in practice often depends upon stability considerations.

2.7.2 Location of the machinery space

Every mechanically propelled merchant ship must have a machinery space which is large enough to contain the main engine, its associated auxiliaries and access trunks, without encroaching excessively upon valuable cargo space. Until the 1950s the usual position for the machinery space in dry cargo vessels (other than certain coasters and colliers) was at, or slightly abaft, amidships. Here it was conveniently placed alongside coal bunkers and allowed the ship to maintain a reasonable trim (the difference between forward and after drafts) when in ballast. But with the increase in the size of bulk cargo shipments and the almost universal adoption of mechanized cargo handling, stowage and utilization, it has become clear that the midships engine-room arrangement has several disadvantages. The most important of these are the need for a tunnel enclosing the propeller shaft which obstructs the after holds and the unproductive use of the parallel middle body of the hull for the engine instead of for cargo, to which it is ideally suited.

In most ships today the usual practice is to locate the engine room as far aft as possible, thereby eliminating the shaft tunnel and freeing the middle of the ship for cargo. Bulk carriers and other full-form vessels invariably have all their holds forward of the engine room. Container ships, however, and some other fast cargo vessels often have a hold abaft the engine room because their after lines are so fine that there is insufficient space for the machinery at the extreme aft end of the hull.

Part of the cargo space in Ro-Ro vessels is usually situated over the engine room. Thus, while the longitudinal position of the engine within the hull may not be critical, the height of the machinery compartment is often limited. This is one of the considerations which has led to the widespread adoption in Ro-Ro vessels of compact sets of medium-speed diesel engines connected to the propeller via gearing instead of directly, as are the much larger slow-speed diesels.

2.7.3 Deck layout

2.7.3.1 *Weather deck*
The weather deck is the uppermost boundary of the hull which, besides protecting its interior against the elements, provides a platform for cargo and machinery such as cranes and winches. It also allows the crew access from one end of the ship to the other. The height of the weather deck at the centre line above the keel usually increases towards the ends of the ship (known as sheer) although in some modern vessels this distance is kept constant.

An opening, protected by trunking within the superstructure, is cut in the weather deck to allow access to the machinery space, and in vertical loading ships further access openings, protected by hatch covers, are cut in the way of

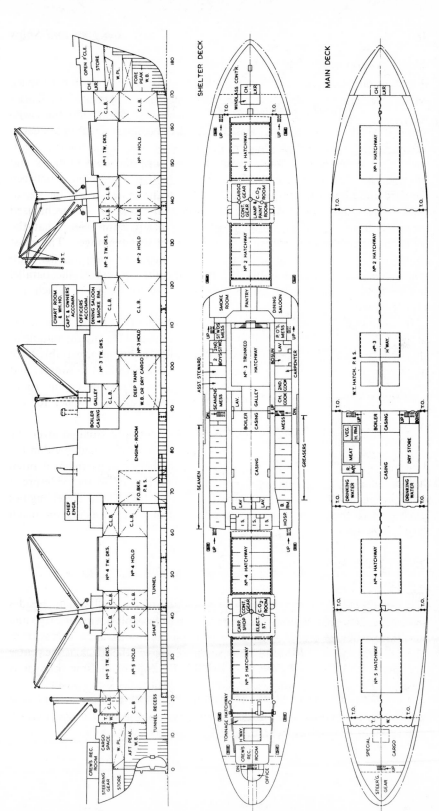

Fig. 2.9. General arrangement of the 10 500 tdw *Essex Trader*, a general cargo/tramp built in 1958. This vessel has many of the features of the traditional cargo ship. Nos. 2 and 3 holds are continuous and are served by hatchways (trunked at No. 3) separated by the bridge structure. Derricks and winches are raised up on mast houses. Note the tonnage hatchway aft and the coamings to the tween-deck hatchways.

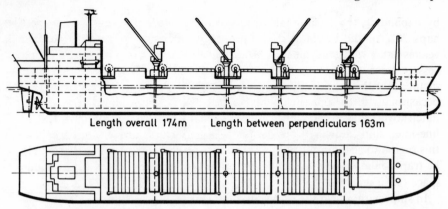

Length overall 174m   Length between perpendiculars 163m

Fig. 2.10. Profile and deck layout of a modern, wide hatch, multi-purpose cargo vessel, fitted with roll-stowing hatch covers on the weather deck and sliding covers in the tween-deck.

each hold. In the simplest and most common arrangement, all the hatchways are rectangular; they have the same width (although not necessarily the same length) and are symmetrically disposed about the centre line. Very long holds may have more than one hatchway, but a single hatchway rarely serves more than one hold.

During recent years a number of variants on this basic configuration have become popular. Their purpose is to expose the greatest possible hold area so that cargo can be lifted directly into or out of its stowage position with a minimum of horizontal movement. This means that the total hatchway area throughout the ship must be as large as possible and in bulk carriers (especially those intended for the carriage of forest products) it is achieved

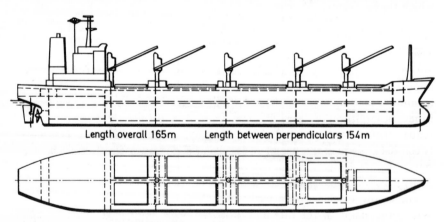

Length overall 165m   Length between perpendiculars 154m

Fig. 2.11. Profile and deck layout of the *Frontier* class standard multi-purpose cargo vessel. Note the twin hatch arrangement at all hatchways except No.1.

by expanding the traditional arrangement, where hatchways occupy perhaps 20% of the available deck area, to its maximum extent. In modern vessels hatch width may be up to 80% of the overall breadth. In general cargo vessels, however, considerations of strength and tween-deck stowage often dictate that the hatchway serving a particular hold be composed of two, or occasionally three, separate openings side by side, each equipped with its own hatch cover. Multiple hatchways are not usually found at the foremost hold in fine-lined ships, however, because the weather deck is rarely wide enough in this region to accommodate them. The forward hatchways may occasionally be trapezoidal or even 'T' shaped in order to make best use of the available space (see Fig. 2.14).

In ships without cargo-handling equipment such as certain container ships and bulk carriers, there is usually very little clear deck space at the ends of the hatchways, but in other vessels, the longitudinal separation of hatchways must be sufficient to accommodate cranes or derricks and winches. Nowadays, winches are often located on raised platforms instead of directly on the deck as was common in the past. This protects winches from sea-water damage, provides the winch drivers with a better view of the hold than would otherwise be possible and, if necessary, allows the hatch covers to be stowed beneath the platforms. Winches may be used to open and close hatches. Cranes are often supported on tall pillars.

A few vessels, notably forest products carriers and certain container ships, are equipped with travelling-portal gantry cranes which straddle the full width of the ship and move along the weather deck on rails situated outboard of the hatch coamings (see Fig. 7.4). Such gantries can only be effectively employed in full form ships with extensive parallel middle body, or in fine form ships with knuckles or considerable flare forward, because the rails on which they run must be parallel.

2.7.3.2 Tween-decks

Tween-decks (between decks or intermediate decks below the weather deck) are found mainly in general cargo and refrigerated vessels. They have two important operational functions. Firstly, they provide 'shelf space' which facilitates loading and discharging and stowage of cargo for multi-port itineraries, although large hatchways and multiple hatchways tend to reduce the amount of space which is available for this purpose. Secondly, tween-decks reduce the height of the column of cargo which must be supported by the lowest tiers in the hold – this is particularly important when easily crushed break-bulk packages such as cartons are carried.

The layout of each tween-deck closely follows that of the weather deck. Where the shape of the hull allows, tween-deck hatchways have the same dimensions as those situated immediately above and in modern vessels, tween-deck hatch covers are flush fitting to provide a smooth and continuous surface for the operation of fork lift trucks. The height of the tween-deck space and the clearance under the weather deck hatch coamings are also adjusted to permit the passage of fork lift trucks. In many ships tween-deck height is determined by the dimensions of frequently-carried packages, e.g. containers or palletized cartons of fruit.

2.7.3.3 *Vehicle decks*

Vehicle decks provide the stowage area for cargo in Ro-Ro vessels. They are approached from the quay by way of ramps and doors at the bow or stern or at the side (shell) and usually extend over practically the whole of the ship's length, although in some container ships with Ro-Ro capacity, the vehicle deck may occupy much less than this. The main vehicle deck invariably extends from one side of the ship to the other and to facilitate the passage of vehicles such as trailers or fork lift trucks, it must be as clear of obstructions as possible. Often there are no transverse bulkheads throughout its entire length and even when these are fitted, they have large openings, often with steel doors, to allow easy access for vehicles from one compartment to another.

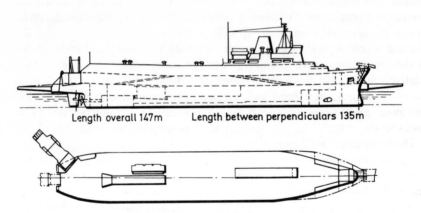

Length overall 147m   Length between perpendiculars 135m

Fig. 2.12. Profile and vehicle deck layout of freight Ro-Ro. Note the bow and stern axial ramps and the quarter ramp situated on the port side. Internal ramps and elevators provide access to the three main decks.

The height of the vehicle deck space depends upon the purpose for which a particular Ro-Ro vessel is constructed. In some cases it may be high enough for containers to be stowed in two tiers, while in specialized car carriers, each vehicle deck space may be just high enough to accommodate private automobiles. Portable mezzanine decks are often used to divide high vehicle deck spaces when this is warranted by the nature of the cargo being carried. Cargo is transferred from one deck to another by means of internal ramps and elevators. Such considerations are discussed more fully in Chapters 6 and 8.

Many Ro-Ro vessels can be used for military purposes. Vehicle decks are often suitable for heavy wheeled or tracked vehicles, which can be discharged onto (makeshift) quays or even into lighters. A few Ro-Ro vessels are designed to beach. Detailed discussion of the military potential of Ro-Ro cargo vessels is beyond the scope of this book, however, as too are the various types of access equipment fitted in warships e.g. elevators in aircraft carriers.

2.8 Dry cargo ship types

2.8.1 Ore and bulk carriers

The origins of modern ore and bulk carriers are to be found in the many single-deck colliers which began to enter service around the turn of the century, but most of their development has occurred since the 1950s with the growth of steel production and the need for ships to transport large quantities of iron ore efficiently.

Iron ore occupies very little space in a ship. In conventional holds it forms shallow piles unsuited to grab discharge. Thus, in many ore carriers longitudinal bulkheads are fitted in every hold in the way of the hatch side coamings, to enclose just enough volume to accommodate a full ore cargo (as shown in Fig. 7.3). This results in a deeper stow which is more readily discharged by grabs. It also improves seakindliness. The wing spaces, outboard of the longitudinal bulkheads, are used for water ballast. Because of their relatively small hold volume, however, ore carriers arranged in this manner are not suited to the carriage of less dense bulk commodities such as coal and grain, and the bulk carrier has been developed to overcome their drawbacks in this respect.

The large wing tanks of ore carriers are replaced in bulk carriers by sets of wedge shaped upper and lower wing water ballast tanks, giving the holds a hoppered cross section as shown in Fig. 7.1. This facilitates loading, stowing and trimming and the upper tanks effectively reduce the cargo free surface area* too – an important consideration in the carriage of grain. Moreover the total hold volume is greater in bulk carriers than in ore carriers of comparable dimensions, thereby allowing them to load to their marks (maximum draft) with fairly low-density commodities. In some vessels, the upper wing tanks can be used to provide additional cargo space when required. Bulk carriers can carry iron ore and many are strengthened so that full cargoes may be loaded using alternate holds only.

2.8.2 Combination Carriers

The flexibility of bulk cargo ships has been further increased with the rapid

---

*A free surface allows the commodity to move in a seaway, thereby reducing the ship's stability.

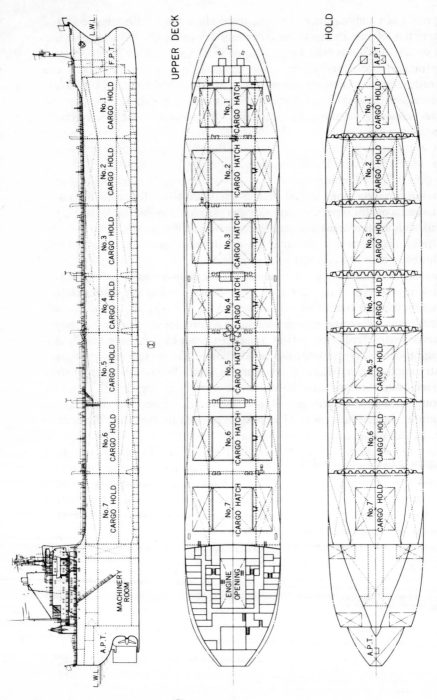

Fig. 2.13. Profile and deck layout of the 66 000 tdw gearless bulk carrier *Ingwi*. This ship is typical of the 'Panamax' type and has side-rolling covers to its seven holds.

development of combination carriers in the early 1960s. These combine the characteristics of ore carriers, bulk carriers and tankers and are usually known by the acronyms O/O (ore/oil), and OBO (ore/bulk/oil).

Some tankers occasionally carry grain if market conditions are right, and it is to take advantage of such opportunities that combination carriers have been developed. Tankers are not ideal for dry cargo, not even grain, because they are primarily intended for the carriage of liquids and have small hatches and no inner bottom. In contrast, purpose-built combination carriers are essentially bulk or ore carriers that can be adapted for liquid cargoes – a rather easier transformation.

The general arrangement of ore/oil carriers is similar to that of ore carriers, but the wing tanks are used for cargo oil as well as for water ballast. In most vessels of this type, the ore holds are used as centre oil tanks and are equipped with all the necessary piping and oil-tight hatches.

Ore/bulk/oil carriers are similar in layout to bulk carriers, with all their holds doubling as oil tanks. When oil is carried, its free surface area is kept small, usually little more than the area of the hatchway, by the upper wing-tank plating, thereby ensuring adequate stability. Hatchway width rarely exceeds 50% of the beam of such vessels.

The pumping and piping arrangements in combination carriers are usually very simple, with direct line systems linked to a pump-room aft being the most common. Heating coils must usually be removed from compartments in which dry cargo is carried, but arrangements with coils next to the hopper plating are quite common.

Although combination carriers can in principle change easily from dry to liquid trades, in practice many of them stay in one or the other for long periods, the majority in the oil trades.

2.8.3 Multi-purpose and general cargo ships

The multi-deck general cargo ship (or freighter) is the traditional 'maid of all work'. Many of its former tasks have now been taken over by container and other unit-load vessels, but a large number of these vessels are still employed either supplementing liner services or in some bulk trades. Many of them have been built as replacements for the standard Second World War Liberty ships.

There are now numerous standard designs of ships which fall into this category, including the well known SD 14 and Freedom types. Most are less than 20 000 tdw and are offered in a variety of versions to suit particular owners' requirements. Although these are not intended to serve specifically as container ships, the standard range of container sizes exerts a considerable influence over their dimensions and arrangement, particularly hatch and deck layout, as discussed in Section 8.1.1. General cargo vessels employed in liner services are frequently called upon to carry containers.

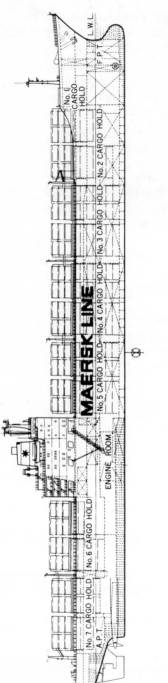

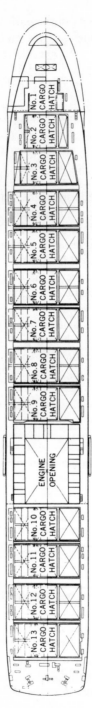

TANKS

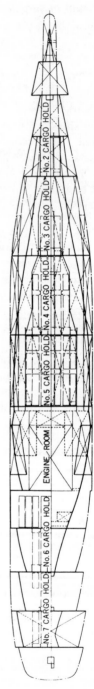

Fig. 2.14. Profile and deck layout of the 27-knot 32 000 tdw cellular container ship *Svendborg Maersk*. Note the shape of the forward hatches, particularly the trapezoidal No. 3 wings, which are all fitted with pontoon covers. This ship can carry 1800 TEU. Containers carried on deck can be either 20 ft or 40 ft long.

2.8.4 Cellular container ships

The cellular container ship is the vessel now most commonly used to transport containers across the seas. The first vessels of this type entered service some twenty years ago and were mostly converted cargo liners and tankers. Their most obvious distinguishing feature is the arrangement of their holds. Unlike the general cargo vessels which they replaced on many routes, with their tween-decks, relatively large holds and small hatches, container ships are single deckers and their holds and hatchways occupy the same area, the dimensions of which are determined by the container sizes. Containers are stowed in the holds in columns supported by vertical steel cell guides – hence the 'cellular' configuration (see Fig. 2.14).

Most container ships are capable of carrying a significant number of containers on deck, which in practice means supported on the hatches. Ships

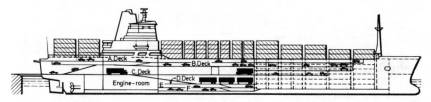

Fig. 2.15. Inboard profile of an ACL G2 (second generation) Ro-Ro container ship. Note the long ramps to spaces below the freeboard (C) deck. Cars can be carried outboard of the cellular container spaces forward. A more detailed arrangement is shown in Fig. 8.13, while Fig. 2.4 shows one of the six vessels in service.

with transom sterns are becoming increasingly common because this provides a roughly rectangular after weather deck area which is more suited to the stowage of containers than the more traditional rounded weather deck shape. Total carrying capacity of most container ships now in service is less than 3000 TEUs (Twenty Foot Equivalent Units).*

The service speed of container ships is higher than that of general cargo liners, usually more than 20 knots. Some exceptional vessels have been built capable of more than 30 knots, but most vessels which have been ordered since the 1973 oil crisis have been slower.

2.8.5 Ro-Ro unit-load vessels

There are now a variety of unit-load container and Ro-Ro vessels. Many combine the attributes of container ships and Ro-Ro ships in extremely ingenious ways. In one arrangement the after end of the hull is fitted out with vehicle decks for Ro-Ro cargo, while the forward end has holds for

---

* All container spaces in a ship can be expressed as 20 ft equivalent spaces, e.g. one 40 ft container = 2 TEUs.

containers as in a cellular container ship. Containers are stowed on the weather deck too. Ships of this type have been operating successfully over a number of transoceanic routes for some years and of particular note are those of the Atlantic Container Lines (ACL), illustrated in Fig. 2.15.

2.8.6 Ro-Ro vessels

Until the early 1970s pure Ro-Ro vessels were believed to be suitable only for short sea trades and many of them were operated as passenger/vehicle ferries (as shown in Fig. 2.16). Most were therefore fairly small. Larger Ro-Ro vessels have recently been found suitable for a number of longer routes, notably to the Middle East and West Africa, where inadequate port

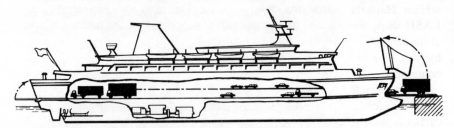

Fig. 2.16. The 6600 GRT Baltic ferry *Visby*. This illustration shows the bow visor in the open position. 1500 passengers together with up to 300 cars stowed on fixed and portable decks can be carried.

facilities have been less of a handicap to Ro-Ros than to conventional general cargo ships. Not only can they load and discharge quickly, but the angled ramps with which many of them are equipped make them independent of quayside handling equipment.

2.8.7 Vehicle carriers

With the growth in export trades in motor cars the number of specialized vehicle carriers in service has increased rapidly. Many of these vessels have capacities in excess of 3000 European size motor cars, stowed on several decks (as shown in Fig. 2.17).

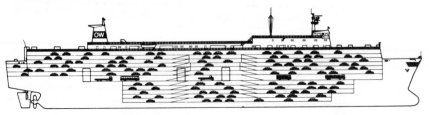

Fig. 2.17. Profile of *Don Juan*, a large 19-knot purpose built car carrier with a capacity of about 4300 cars stowed in 12 decks. Side doors and ramps provide external access at freeboard deck level; this deck has a greater height for commercial vehicles.

2.8.8. Barge carriers

There are now in service a number of vessels designed to carry standard barges, each with a capacity of several hundred tonnes. In one design – the LASH (Lighter Aboard SHip) concept – loaded rectangular barges are towed to the anchored carrier which loads them over the stern by means of a massive mobile gantry. Once on board they are transported forward by the gantry and lowered into their designated stowage position within the hull by way of large weather deck hatches. All the ship's superstructures, apart from twin narrow funnels, are forward of the cargo spaces so as not to obstruct the movements of the gantry.

An alternative arrangement – the 'Seabee' concept – is also established in service. Here the rectangular lighters, which are larger than those used by LASH ships, are raised clear of the water by a hydraulic elevator situated at the carrier's stern. They are then transported on rails forward into the hull, by way of a stern opening. The weather deck area may also be used for containers, loaded and discharged by crane in the usual manner.

A third, smaller barge-carrying system known as BACAT (Barge Aboard CATamaran) has also been introduced. In this arrangement a catamaran carrier is used to carry barges between its hulls, and on deck.

More detailed descriptions of these vessels and their cargo access equipment are given in Chapters 7 and 8.

CHAPTER THREE

The Influence of Cargo Access Equipment on Ships' Performance

Summary

This chapter identifies the principal elements of the time spent in port, and shows that about half of quay-to-quay transport costs are attributable to such time. Delays in port directly attributable to cargo access equipment now form a very small proportion of total delays. Correct choice of access equipment can considerably reduce overall port times, with benefits to both shipper and shipowner.

3.1 The cost of time in port

Merchant ships earn their living by carrying cargoes across the seas, and so the time that they spend in port should be kept to a minimum. Faster turnrounds can be achieved by increasing cargo handling rates or by reducing time spent unproductively, for example preparing for cargo work or awaiting better weather. Efficient cargo access equipment contributes to both of these areas by facilitating cargo handling operations, and by cutting the time needed to open up and batten down cargo spaces.

The length of time that a ship stays in port is influenced also by its type and size, the nature and quantity of cargo it handles and the facilities of the port. Table 3.1 shows typical operational profiles for seven ship types derived by the authors from actual service records. It indicates that most types of dry cargo ship spend between 25 and 50% of each round trip in port.

A considerable proportion of the annual costs of operating a ship are incurred in port. This not only includes direct costs like cargo handling, but also costs such as crew wages, which must be borne in port as well as at sea. Tables 3.2 and 3.3 show the breakdown total annual costs in 1975–76 for two of the ships in Table 3.1, derived from analysis of actual British ship operating costs. The tables indicate that about half of the total quay-to-quay transport cost is attributable to time in port. In many bulk trades, however, the shipowner does not pay directly for cargo handling costs, but in such cases

Table 3.1. Typical ship voyages and port times

Ship type (Typical cargoes)	Dead-weight tonnes	Speed knots	Round trip nautical miles	Round trips per annum	Number of ports of call	Days per call	Port days per round trip	Sea days	Total days per round trip	Per-centage port time
Fast cargo liner (manufactured goods)	17 000	18	17 000	4·2	11	4	44	39	83	53
Break-bulk freighter (manufactures, steel, grain)	15 000	15	15 000	5·0	5	5·5	28	42	70	40
Large container ship (2300 20 ft containers)	36 000	23	12 000	11·3	6	1·5	9	22	31	29
Large bulk carrier (iron ore, coal)	110 000	15·5	10 000	10·0	2	4	8	27	35	23
Small bulk carrier (grain, forest products)	25 000	14·5	11 000	7·6	4	3·5	14	32	46	31
Ro-Ro freight ferry (90 trailers or equivalent)	5 000	17	800	116	2	0·5	1	2	3	33
Coaster (coal, grain, timber)	3 000	12	1 400	39	2	2	4	5	9	45

Note: assumed each ship in service about 350 days per annum

Table 3.2. Annual costs for fast cargo liner
(1975–76 prices)

Ship and cargo		
Deadweight, tonnes	17 000	
Bale capacity, cubic metres	23 000	
Speed, knots	18	
Round trips per annum	4·2	
Cargo carried per year, allowing for high stowage factor cargoes, tonnes	72 000	

Annual costs	£	%
Crew wages and benefits	220 000	8·6
Victualling and stores	50 000	2·0
Maintenance and repair	130 000	5·1
Insurance	50 000	2·0
Administration and sundries	50 000	2·0
Total 'daily' running costs	500 000	19·7
Port charges and canal dues	180 000	7·1
Fuel costs	470 000	18·4
Total non-cargo voyage costs	650 000	25·5
Cargo loading and discharging	690 000	27·1
Cargo claims and commissions (7% of freight)	180 000	7·1
Total cargo costs (£12 per tonne)	870 000	34·2
Total operating costs	2 020 000	79·5
Capital charges, 1971-built ship*	525 000	20·5
Total costs to be recovered from freight earnings	2 545 000	100·0

Unit costs	£	%
Quay-to-quay freighting cost per tonne	35·3	62·7
Typical inland haulage costs per tonne	15·0	26·6
Typical miscellaneous costs (cargo insurance, lightering, handling within port, customs, documentation etc.)	6·0	10·7
Total transportation costs per tonne	56·3	100·0

Total annual ship costs incurred in port can be estimated as: port charges, plus cargo handling costs, plus the proportion of daily running costs and capital charges attributable to time in port, i.e.

180 000 + 690 000 + 53% of (500 000 + 525 000)   = £1 413 000
Total annual quay-to-quay costs   = £2 545 000
Percentage incurred in port = 1 413 000/2 545 000 = 55·5%

---

*Typical cost of an existing cargo liner £3·5 M; if 1977 newbuilding contract, cost increased about 120%, but by the time of delivery, other operating costs will have risen too

Table 3.3. Annual costs for large bulk carrier
(1975–76 prices)

Ship and cargo		
Deadweight, tonnes		110 000
Speed, knots		15·5
Round trips per annum		10·0
Cargo carried per year allowing for ballast voyages, tonnes		1 130 000

Annual costs	£	%
Crew wages and benefits	230 000	7·4
Victualling and stores	60 000	1·9
Maintenance and repair	210 000	6·8
Insurance	170 000	5·5
Administration and sundries	60 000	1·9
Total 'daily' running costs	730 000	23·5
Port charges	198 000	6·4
Fuel costs	970 000	31·3
Brokerage and commissions (2½% of freight)	77 000	2·5
Total voyage costs	1 245 000	40·2
Total operating costs	1 975 000	63·7
Capital charges, 1971-built ship*	1 125 000	36·3
Total costs to be recovered from freight earnings	3 100 000	100·0

Unit costs	£	%
Sea freighting cost per tonne cargo	2·74	66·2
Typical loading cost, per tonne	0·60	14·5
Typical discharging cost, per tonne	0·80	19·3
Total transportation cost, stockpile to stockpile	4·14	100·0

Total annual ship costs incurred in port can be estimated as: port charges, plus the proportion of daily running costs and capital charges attributable to time in port, plus cargo handling costs, i.e.

$198\,000 + 23\%$ of $(730\,000 + 1\,125\,000) + (0·60 + 0·80) \times 1\,130\,000 = £2\,207\,000$

Total annual quay-to-quay costs $= 4·14 \times 1\,130\,000$ $\qquad = £4\,678\,000$

Percentage incurred in port $= 2\,207\,000/4\,678\,000$ $\qquad = 47·2\%$

---

*Typical cost of an existing bulk carrier £7.5M; if 1977 newbuilding contract, cost increased about 70% but by the time of delivery, other operating costs will have risen too the freight rate that he receives is adjusted by an appropriate amount. Nevertheless, cargo handling charges must be included in any comparison of alternative transportation methods involving improved designs of ship.

3.2 Delays in port

Until the 1950s, dry cargo handling methods had barely changed for decades. Cargo access equipment manufacturers helped to expedite ships in port by improving the design of hatch covers, and making possible savings of both time and cost. A traditional 'beams and boards' hatch in a typical cargo vessel could take a gang of ten men up to an hour to open or to close when making the ship ready for sea. This is an unproductive time cost made up not only of the wages of the labour force, but also the loss of earnings for the ship during the period. By comparison, an automatic mechanical hatch cover can be opened by one man in about two minutes, or two men can open a hatch operated by wires and winches in about 10–15 minutes (less if the wires are already rigged).

By 1965, a large proportion of ships had mechanical weather deck covers; delays due to opening and closing such hatches had fallen to only 2·7% of cargo working hours, according to a major survey of general cargo operations carried out in Dutch ports [3.1]. Thus during a typical 8½-hour shift, time lost through this cause amounted to only about 14 minutes. The corresponding figure for tween-deck hatch covers was 1·3%, showing that operating time for hatch covers had ceased to be a major problem following the widespread adoption of improved access equipment. With present-day designs, the time lost has fallen even further. Table 3.4 shows other results of delays during cargo working hours from the same survey.

Table 3.4. Breakdown of delays during cargo working hours

	Delays as a percentage of working hours	*Average minutes per shift*
Opening and closing all hatches	4·0	21
Refreshment breaks	3·7	19
Weather delays	3·0	15
Setting up or shifting cargo gear, lighters etc.	2·1	11
Waiting for cargo, stevedores or lighters	2·0	10
Other delays (dunnage, re-stowing cargo etc.)	4·7	24
	19·5	100

Note: Figures relate to average of loading and discharging general cargo during a normal 8½-hour shift, which excludes a ¾-hour meal break.

Table 3.5 shows a breakdown of total time in port, derived from an analysis of the performance of a number of individual cargo liners. Those for other types of ship with quicker turnrounds, e.g. bulk carriers or unit load ships, do not differ as much as might be expected because, while their cargo-handling operations proceed more rapidly and they usually work more shifts, their shorter port times mean that nearly constant delays, such as waiting for berths or preparing for sea, have a proportionately greater effect.

Table 3.5. Breakdown of time in port for typical break-bulk general cargo vessel

Item	Percentage	
Cargo being worked	32	
Delays during working shift (Table 3.4)	8	
Stevedores available for work		40
Waiting for berth	5	
Waiting at start and finish of cargo operations, waiting for stevedores	10	
Nights and regular meal breaks	31	
Weekends and holidays	8	
Weather delays, shifting berth, repairs	6	
Total non-working hours		60
Total time in port		100

Note: For a typical port call of about 4 days, the percentages may also be read as hours.

Operating times for items of access equipment other than hatch covers, e.g. ramps and stern doors on Ro-Ros, are generally only a few minutes and cause virtually no delay to cargo operations. The major contribution that improved access equipment has to offer now is in providing a means of expediting the cargo work itself, particularly in general cargo vessels. In the 1960s, a major contribution was the provision of flush tween-deck hatch covers, which permitted fork lift trucks to operate freely, unimpeded by hatch coamings. Much of the work in the holds consisted of stowing and retrieving cargo from the wings (i.e. outside the line of the hatchway opening) and thus significantly higher cargo handling rates became possible after flush tween-deck covers came into general use. The manual effort required by stevedores was reduced as a result of the introduction of such methods and was an important consideration leading to their widespread adoption.

Additional advantages were the reduction in damage caused by dragging items of cargo over the coamings, and an increase in usable tween-deck cargo volume. Removal of the 228 mm (9 in) high coamings in a typical large cargo liner increased capacity by about 100 m³ (3500 ft³) per deck, corresponding to about 90 freight tons additional payload.

3.3 Benefits of reduced time in port

The benefits of reduced time in port vary for different types of ship and also for the types of charter under which a ship may operate. For bulk cargo vessels, which usually carry homogeneous cargoes like grain or iron ore between two ports, the benefit of reduced time in port appears as a shortened round voyage cycle time, permitting more cargo to be carried in a year; in most bulk trades, there is usually additional cargo available to fill the extra capacity so created.

A ship trading on voyage charter is contracted to carry cargo between specified ports or areas and, as soon as the cargo has been delivered, the ship is free to seek further employment. If the port time is less than the 'lay days' which have been agreed in the charter party (the time allocated for loading and/or discharging), the shipowner must reimburse the charterer by paying 'despatch', at a predetermined sum per day. Conversely, a longer port time results in the shipowner receiving 'demurrage' to compensate for having his ship delayed. However, the normal rates of demurrage and despatch are such that all parties have an incentive to turn the ship around as quickly as possible.

Reduction in port time has little effect on port charges which include towage fees, conservancy dues, dock rents and the like, as these are usually charged per gross or net registered ton (see Section 4.2.1) or per tonne of cargo worked. A small benefit would however be gained at those ports where dock rents have a large 'per diem' element. Similarly, cargo handling charges are usually at an agreed rate per tonne and are not raised when port turnround time is less than anticipated unless this has been achieved by working overtime.

In many bulk trades, cargo handling charges are to the charterer's rather than the shipowner's account, as quayside cargo handling equipment is normally used at specialized loading or discharging terminals. The benefit thus accrues to the shipowner as increased annual earnings from more voyages, and to the charterer as despatch. If reductions in port time are consistently achieved for the majority of ships on a trade, the customary lay days may be reduced, resulting in lower freight rates, to the advantage of the final customer, say the housewife, who buys the loaf of bread made from the grain being shipped.

For time chartered vessels, which are at the disposal of the charterer for an agreed period of time – from a few months up to 10 years or more – the direct benefit of reduced port time accrues to the charterer since he, not the shipowner, is responsible for arranging the individual voyages and ports of call for the ship. He may perhaps redeliver the ship earlier (within agreed limits) than expected, so reducing the total time charter hire payment, or alternatively he may be able to squeeze in an additional voyage within the maximum agreed charter period.

If a shipowner can demonstrate to the charterer in advance that the ship's equipment is such that better performance and less port time can be expected, compared with that offered by other ships, he may be able to negotiate a higher charter rate. In poor markets, such a gain may not be possible; the potential advantage to the owner then is to have his better equipped ship chartered rather than a competitor's. The penalty for failure in such circumstances may be to have to lay up his ship or accept an even less attractive charter.

For owner-operated ships – either liners on scheduled services, or industrial carriers providing transport for their own products and raw materials (e.g. steel companies) – the situation is rather different. On liner services, only limited amounts of cargo are likely to be available. Furthermore liners usually operate within a Conference, which usually regulates sailings and freight rates in consultation with shippers. In these circumstances, reductions in port time may not necessarily allow more cargo to be carried each year or more voyages to be made, unless the existing capacity is too small. However, if the reductions are predictable and sustained, e.g. by substituting Ro-Ro vessels for less productive tonnage, an improved service and/or shorter voyage time may be possible, offering benefits which outweigh the additional investment.

Occasional reductions in port time are relatively less beneficial, as it may not be possible to take advantage of them without upsetting schedules. Where some flexibility is possible, reductions in port time can show benefits such as an earlier arrival time at the next port to catch a tide or berth or the start of the stevedores' shifts, or to avoid a chain of costly consequences like overtime working for stevedores and demurrage on lighters or road transport. Moreover, time saved can permit slower steaming on the next leg of the voyage, so reducing fuel consumption and cost.

In practice, the proportion of voyage time spent in port by break-bulk cargo liners has not fallen very much in recent years, because much of the benefit of faster cargo handling operations and improved access equipment has been offset by increasing the size of the ship to obtain economies of scale. Consequently cargo liners of 15 000–20 000 tdw now spend no longer in port than did the previous 10 000 tonners, but reap the benefits of size in the form of less than proportionate increase in fuel consumption, crew number and building cost.

In the bulk trades, improved cargo access equipment has facilitated the development of ships like the open 'all-hatch' forest products carriers (see Section 7.1.6) which by their nature have improved transportation efficiency.

The actual monetary benefit of reduced port time depends on whether long-run or short-run costs are being considered. Goss [3.2] discusses the effect of long-run costs, identifying these as the daily running costs of the ship, including crew costs, stores, maintenance and repair, insurance and administration (i.e. excluding any voyage costs like fuel, port or cargo handling costs), plus capital charges, i.e. the sums of money needed to recover the initial investment in the ship together with an appropriate rate of return on capital, converted to an annual or daily basis. In the case of a time charter, where a ship and its crew are hired for a period, the above costs are exactly those which a shipowner must meet out of hire income, so that it is possible to obtain an indication of the daily cost of any ship's time from a consideration of the corresponding long-term time charter freight rates. Such long-term rates are set more by the actual first cost of the ship and its anticipated running costs, than by short-term fluctuations in the freight markets due to variations in the supply and demand for ships. It is therefore possible to quantify the benefits in the long run of savings in port time by converting long-term time charter rates to equivalent daily figures. Time charter rates for a ship are sometimes expressed as so many dollars per day, but more usually per unit of capacity per month. The unit of capacity is often the ton deadweight, but could be the cubic foot for refrigerated ships, the cubic metre for liquefied gas carriers, the twenty foot equivalent unit (TEU) for container ships, or some other measure appropriate to a particular ship. Thus a rate of (say) $10·00 per ton deadweight per 30-day month for a 15 000 tdw ship is equivalent to

$$10 \times 15\,000/30 = \$5000 \text{ per day*}$$

In the short run, market rates may be above, or more often below, the level needed to cover running costs and provide an adequate return on capital. It is then not so easy to quantify the cost of saving time in port except by considering a specific ship, a specific trade, a specific port, and a specific market level, where all the individual costs can be identified, including such detailed items as overtime rates. Fortunately, the problem of selecting the best cargo access equipment for a ship is not influenced by short-run costs, but by long-run costs, as decisions have to be made at the design stage which will largely determine the ship's performance over its entire life.

Thus, the greatest benefit to the shipowner is obtained when improved access equipment is planned in consultation with equipment manufacturers, shipbuilders and port authorities before a ship is built. This enables not only the whole shipping operation to be closely geared to the capabilities of the ships and their equipment, but also the detailed design of the ship to be tailored to the requirements of the cargo, its handling and stowage on board, and the capabilities of the ports to be served.

* Freight rates are nowadays usually expressed in US dollars.

Table 3.6. Potential benefit of 10% reduction of port time

Vessel	Capacity	Typical round trip time spent in port	Typical newbuilding cost, million dollars	Timecharter rate, dollars per unit capacity per month	Annual benefit of reduced port time, dollars	Capitalized port benefit, dollars (D)	Percentage of newbuilding cost
Large bulk carrier	110 000 tdw	20% (8 days)	20–25	(A) 3·75	80 000	530 000	2·3
				(B) 1·20	25 000	170 000	0·8
Fast container ship	2300 TEU (36 000 tdw)	25% (9 days)	50–70	(A) 400	200 000	1 300 000	2·2
				(C)			
Short sea Ro-Ro ferry	110 12 m trailers	40% (1 day)	10–14	(A) 2040	105 000	700 000	5·8
Break-bulk freighter	(6000 tdw) 15 000 tdw	40% (28 days)	7–10	(B) 3000	155 000	1 050 000	8·8
				(A) 10.20	75 000	500 000	6·1
Short sea coaster	3000 tdw	48% (4 days)	2–3	(B) 8·00	58 000	390 000	4·8
				(A) 21·00	35 000	230 000	9·2
				(B) 17·00	28 000	180 000	7·2

(A) Indicates long-run cost levels at 1976–77 prices.
(B) 1976–77 average market rates, which were low for bulk vessels.
(C) No regular charter market for such ships.
(D) Capitalized benefit based on Capital Recovery Factor of about 15%, corresponding to a rate of return of about 14% over 20 years.
  Annual benefit = 0·15 × capitalized benefit.

Table 3.6 shows the potential economic advantage for several typical ship types, resulting from reduced port time, as estimated by the authors. The savings have been estimated by comparing each alternative 'with' and 'without' the reduction; that is, full advantage has been taken of the assumed 10% reduction in port time, to shorten the overall voyage cycle time and to increase the annual quantities of cargo carried. The savings are directly proportional to the reduction in port time. Thus the benefit from a 1% reduction in port time can be obtained by dividing the figures in Table 3.6 by 10. While the precise answers depend upon the detailed assumptions made about voyage length, ship cost etc., it is clear that the financial benefit can be substantial. Thus an appreciable additional investment may be worthwhile to achieve such gains, ranging from a minimum of about 1% of the new-building cost for a large bulk carrier in poor markets, to a maximum of about 9% for a short sea vessel. The equivalent capital sum can amount to $1 million or more, which can purchase a wide range of cargo access equipment.

In essence, access equipment must:

(1) facilitate cargo operations;
(2) be safe and reliable;
(3) be simple and quick to operate;
(4) be economical for its capability.

Chapters 7 and 8 discuss in more detail how such equipment can be selected for any particular ship type.

References

[3.1] Melessen, H. K. J. Managerial Tools for Improving Productivity. *9th Int. Conf. of Int. Cargo Handling Co-ordination Assn*, Gothenburg, 1969.
[3.2] Goss, R. O. The Cost of Ship's Time. In: *Advances in Maritime Economics*, Ch. 4, Cambridge University Press, 1977.

CHAPTER FOUR

General Requirements for Access Equipment

Summary

This Chapter describes the requirements of the 1966 International Load Line Convention, and the rules of maritime regulatory bodies including the Classification Societies, where they affect cargo access equipment, particularly hatch covers. It also discusses operational requirements such as coaming heights, hatch cover stowage, deck openings and drainage, as well as the maintainability of access equipment in general terms. Safety and fabrication are also discussed.

4.1 Regulatory requirements

4.1.1 Rules governing cargo access equipment

Cargo ships exist to transport goods from one port to another and to accomplish this, they must have some means of getting cargo into and out of their holds. Thus ships must have openings either in the weather deck, as in vertically loading vessels, or in the bow, stern or side, as in horizontally loading vessels. These areas are all vulnerable to damage at sea. Thus cargo access equipment must provide access in port while keeping water out during the voyage. One of its prime functions is to ensure the safety of the ship, personnel and cargo. In order to maintain acceptable standards of safety at sea, rules and regulations have been introduced over the years, many of which directly concern access equipment.

4.1.2 1966 International Load Line Convention

Probably the most important regulations that affect access equipment are those arising from the 1966 Load Line Convention, (LLC) [4.1] which replaces the first International Load Line Convention agreed in 1930 for determining freeboard (the height of the main deck above the load waterline). It provides for two categories of ship for freeboard purposes – Type A ships which are those designed to carry only liquid cargoes in bulk and whose tanks have only small access openings, and Type B which includes all other ships. Among Type B ships are certain bulk carriers and ore carriers which are further sub-divided as follows:

(a) Ships of over 100 m in length having steel hatch covers on all exposed hatchways and which can remain afloat in satisfactory equilibrium with any one compartment (excluding the machinery space) flooded. When the length exceeds 225 m, the machinery space is also treated as a floodable compartment. These ships can have 'Table B' freeboards, reduced by 60% of the difference of the freeboards given in Tables B and A, summarized in Table 4.1; such freeboard is referred to as 'B − 60' (spoken as 'B minus 60'). Tables A and B refer to the minimum freeboards laid down by the Convention for Type A and Type B ships.

(b) Ships which can withstand flooding of two adjacent compartments may be assigned 'B − 100' freeboard, virtually equal to Type A ships.

Type B ships which are fitted with portable covers, e.g. beams and boards in Position 1 hatchways (see next section), are required to have their freeboards increased above those given in Table B.

4.1.3 Freeboard

Table 4.1 shows extracts from the full Tables and compares freeboards for different ship lengths as required by the Load Line Conventions of 1930 and 1966. Corrections are made to cater for non-standard length/depth ratios, block coefficient, extent of superstructures etc. The standard freeboard applies in salt water, in designated Summer Zones throughout

Table 4.1. Freeboard tables for the 1930 and 1966 Load Line Conventions

Ship length, metres	Tanker freeboards 1930, mm	Table 'A' freeboards 1966, mm	Steamer freeboards 1930, mm	Table 'B' freeboards 1966, mm
30	–	250	250	250
60	570	573	570	573
90	1015	984	1070	1075
120	1550	1459	1775	1690
150	2170	1968	2540	2315
180	2710	2393	3230	2915
210	(3080)	2705	3810	3430
240	(3370)	2946	(4310)	3880
300	(3830)	3262	(4755)	4630

Note: Figures in brackets show the extension of tabular values as agreed by consultation with other Administrations, (see Section 4.1.4) because the 1930 Convention did not cover ships of these lengths [4.3].

the world; slightly different values apply in Winter or Tropical Zones, in fresh water, or when timber deck cargoes are carried, as laid down in [4.1] and in the corresponding national legislation, (e.g. in the United Kingdom, The Merchant Shipping (Load Line) Rules, 1968 [4.2]). It can be seen that the 1966 LLC permits lesser freeboards, particularly for bigger ships, so increasing maximum draft and therefore cargo deadweight. Table 4.2 is derived from the full Tables and shows the increase in freeboard required for Type B ships having portable covers secured by tarpaulins and battening devices, e.g. beams and boards. Position 1 hatchways are defined as those on exposed freeboard decks, raised quarter-decks, and exposed super-structure decks within the forward quarter length of the ship, while Position 2 hatchways are those on exposed superstructure decks abaft the forward quarter of the ship's length. In this connection a superstructure deck is regarded as the deck immediately above the freeboard deck.

Table 4.2. Freeboard increases for ships with wooden covers

Length, *metres*	*Freeboard increase,* *mm*	*Resultant freeboard,* *mm*
30	50	300
60	50	623
90	50	1125
120	84	1774
150	228	2543
180	313	3228

4.1.4 Freeboard deck

Regulation 3, Paragraph 9 of the 1966 LLC defines the freeboard deck and is quoted here in full as it has an important bearing on all aspects of equipment used for closing cargo openings.

'The freeboard deck is normally the uppermost complete deck exposed to weather and sea, which has permanent means of closing all openings in the weather part thereof, and below which all openings in the side of the ship are fitted with permanent means of watertight closing. In a ship having a discontinuous freeboard deck, the lowest line of the exposed deck and the continuation of that line parallel to the upper part of the deck is taken as the freeboard deck. At the option of the owner and subject to the approval of the Administration, a lower deck may be designated as the freeboard deck provided it is a complete and permanent deck continuous in a fore and aft direction at least between the machinery space and peak bulkheads and continuous athwartships. When this lower deck is stepped the lowest line of the deck and the continuation of that line parallel to the upper part of the deck is taken as the freeboard deck. When a lower deck is designated the freeboard deck, the part of the hull which extends above the freeboard deck is treated as a superstructure so far as concerns the application of the conditions of assignment and the calculation of freeboard. It is from this deck that the freeboard is calculated.'

It should be noted that the requirement for a lower deck to be 'complete and permanent' in order to be designated the freeboard deck, does not mean that any openings need to be weathertight, only that the structure of a 'deck' is present. Tween-deck hatches may be of steel or wood, but are only required to be watertight if fitted to deep tanks or compartments containing water ballast.

The 'Administrations' referred to include the Department of Trade in the United Kingdom, the US Coast Guard in the United States and equivalent bodies in other countries.

4.1.5 Weathertightness and watertightness

Regulation 16(4) of the 1966 LLC which covers the means for securing weathertightness of steel weathertight covers states that

'the means for securing and maintaining weathertightness shall be to the satisfaction of the Administration. The arrangements shall ensure that the tightness can be maintained in any sea conditions, and for this purpose tests for tightness shall be required at the initial survey, and may be required at periodical surveys and at annual inspections or at more frequent intervals.'

Regulation 3(12) defines 'weathertight' as meaning that water will not penetrate into the ship in any sea conditions. 'Watertight' is not defined in the 1966 LLC but is generally regarded as a higher standard than weathertight. It is usually taken to require the closure to be capable of preventing the passage of water through the structure in any direction, under a head of water which generally needs to be defined in each case. This could be the ship's margin line, which is a line drawn at least 76 mm (3 in) below the upper surface of the bulkhead deck at the side of the ship [4.4].

In practice the hatch covers of $B-60$ and $B-100$ ships must be of steel and made weathertight by means of special gasket devices. Covers for Type B ships are usually of this type. Less commonly, hatchways may be made weathertight by means of tarpaulins and battening devices over wooden or portable steel pontoon covers.

4.1.6 Statutory regulations

The 1966 Load Line Convention came into force in 1968 after it had been ratified by the required number of maritime nations. In the United Kingdom the Convention was brought into force by the 1967 Merchant Shipping Act, from which the Merchant Shipping (Load Line) Rules 1968 were made [4.2]; the rules for the construction of ships laid down by Lloyd's Register of Shipping are consistent with the Convention, as are those of other Classification Societies.

4.2 Vertical loading ships

4.2.1 Hatches and tonnage

In 1965 tween-deck hatches on the freeboard deck which were not weather-tight and had no coamings were allowed for the first time. This had implications both for cargo handling efficiency and for cargo access equipment design, thus it is useful to recall briefly the events that led up to this decision.

During the 19th century the method of measuring tonnage was standardized with the introduction of the twin parameters: Gross Registered Tonnage (GRT) and Net Registered Tonnage (NRT). The latter was, and still is, often used as a basis for determining port charges, while dry-docking and towing charges are often related to the former. Various statutory requirements are also related to gross tonnage. Gross tonnage is broadly the volume in cubic metres of under-deck spaces and superstructures (with special allowances for some tween-deck spaces, water ballast compartments, etc.) divided by 2·83 (divided by 100 if the volume is measured in cubic feet), while net tonnage is the residue after approved deductions for machinery spaces and crew accommodation, etc. have been taken from the gross tonnage. Gross tonnage is therefore a broad measure of ship 'size', while net tonnage is a measure of earning capacity.

The tonnage of hatchways [4.8] leading to spaces included in the gross tonnage of a ship is measured by multiplying together their mean length, breadth and depth and dividing by 2·83 (100 for feet units) and deducting from the aggregate of all hatchways 0·5 per cent of the ship's gross tonnage excluding the hatchways. The remainder (if any) of this subtraction is the tonnage of hatchways, customarily referred to as the 'excess of hatchways' and is included in the gross tonnage of the ship.

4.2.2 Tween-deck ships with exempt spaces

The tonnage regulations encouraged the development of designs intended to minimize GRT and NRT. Because tonnage measurements were based on volumes, ships had to pay full dues even when their cargo spaces were only partially loaded. A variety of ship designs were introduced during the 19th century expressly to take advantage of the tonnage regulations, but after 1908 open shelter deck and closed shelter deck ships became the principal alternatives for dry cargo. An open shelter deck ship had a 'tonnage opening' in the weather deck (which was called the shelter deck). The space between the weather deck and the main deck was excluded from the ship's GRT and NRT, provided that the means for closing the tonnage opening and the openings in the bulkheads were of an approved pattern. The second deck was then regarded as the freeboard deck, and all the tiers of tween-decks above as a superstructure. In closed shelter deck ships, the uppermost deck was treated as the freeboard deck. Such vessels were allowed to load to a deeper draft, with greater deadweight, than open shelter deckers, provided that their scantlings were appropriately increased. Ships were often designed to be interchangeable between the 'open' and 'closed' condition, and were then called open/closed shelter deckers (see Fig. 2.9).

Thus ships carrying dense cargoes, where deadweight capacity was important, tended to be closed shelter deck ships, while those carrying light, bulky cargoes which would fill the holds before the ship was down to its marks, were usually open shelter deck ships.

In order to qualify as an open shelter deck ship, it was necessary for the tonnage openings to have 'non-permanent means of closing' (tonnage hatch). Moreover, the second deck had to be provided with weathertight closures since this was now the freeboard deck. For practical purposes, however, this was often regarded as superfluous since the weather deck hatches, including the tonnage opening, were always as weathertight as it was possible to make them.

By 1966 many countries agreed that tonnage openings served no useful purpose and that they were potentially unsafe. Accordingly it was agreed internationally at IMCO (Inter-Governmental Maritime Consultative Organisation) to rationalize the position by permanently covering over or dispensing with the tonnage opening and making provision for ships to be assigned two load lines, corresponding to two tonnages, and preserving the principle of smaller tonnages for ships less deeply loaded. One load line is determined using the weather deck as the freeboard deck, with the tween-deck space included in the tonnage. A second load line is determined using the second deck as the freeboard and tonnage deck; the tween-deck space is regarded as a superstructure, but excluded from the GRT and NRT. The Administration then assigns a 'Tonnage Mark' which is cut into the ship's side at a height corresponding to this second, shallower draft, and provided the mark is not submerged, the smaller values of both GRT and NRT are assumed to apply in assessing port dues etc. If the mark is submerged, the larger values apply. The higher tonnage applies in determining compliance with statutory regulations.

4.2.3 Universal tonnage measurement system

It has long been recognized that existing systems of tonnage measurement contain complexities and illogicalities which have a distorting effect on ship design and operation. After deliberations extending over many years, a new Universal Measurement System for tonnage (UMS) was drawn up by IMCO in 1969. The system is much simpler than any which has been used before. Gross tonnage is calculated directly as the volume of all enclosed spaces in a ship, with no exemptions or deductions, particularly of shelter tween-decks. Net tonnage is largely a function of cargo space volume and number of passengers [4.5].

The 1969 Convention has been accepted by a number of countries. At the time of writing, it is expected to be fully ratified shortly, and come into force with appropriate transitional arrangements within the next few years.

For many ships the new gross and net tonnages are expected to be fairly close to their existing tonnages. The principal exceptions are vessels whose tween-decks are currently exempt. Thus the adoption of UMS is unlikely to have any effect on cargo access equipment in most vessels, although 'paragraph' ships with tween-decks may be affected. Paragraph ships are vessels designed right at the limit where different regulations apply. For example ships of over 1600 GRT require to carry a radio operator and to have higher standards of life saving appliances. Consequently many ships are designed with a GRT of 1599; other limits include 500 tons.

Unless special arrangements are made, paragraph ships are likely to have considerably smaller overall dimensions than those now measured by the present tonnage rules. Thus if the present 'paragraph' tonnages remain, owners will have to decide in future whether to build physically smaller ships with fewer decks, or to disregard tonnage considerations and build whatever size of ship is required for a particular trade.

4.3 Horizontal loading ships

4.3.1 Roll-on/Roll-off ships

It is necessary to consider roll-on/roll-off ships separately because unlike other vessels, their cargo enters and leaves their hull horizontally through openings which are usually below the weather deck and often close to the waterline. Normally the main vehicle deck is the freeboard deck. This means that bow, stern and side doors are above the freeboard deck and hence do not have to be watertight, but must be weathertight, falling into the same category as hatch covers.

4.3.2 Openings in the vehicle deck

It is often necessary to have openings in the main vehicle deck to allow vehicles access to the spaces below. Clearly, no water should enter the lower hull through this opening in the event of the flooding of the vehicle deck. Since this deck is within an enclosed superstructure, there is no statutory requirement, in non-passenger carrying vessels, for openings to have covers. However, prudent operators fit covers to restrict possible flooding, spread of fire, or spillage of fuel and liquid cargoes.

4.3.3 Bow, side and stern doors

Ships on unrestricted service, having a bow visor or bow doors, must have an additional weathertight door inboard, but this is not required for the less exposed stern and side doors. Bow doors must be robustly built and as tight as operational considerations allow, so as to give as much protection to inner doors as practicable. Inner bow doors, side and stern doors must be gasketed and weathertight, with adequate strength and securing arrangements.

4.3.4 Collision bulkheads

All ships are required to have a watertight collision bulkhead situated between about 5% and 8% of the ship's length abaft the forward perpendicular (fore end of the summer load waterline) [4.6]. Where the collision bulkhead extends above the freeboard deck, the extension needs only to be to weathertight standards. Thus where a ship has a bow access, the vehicle ramp in its stowed position can serve as both a weathertight door and part of the collision bulkhead.

4.3.5 Other watertight bulkheads

Every ship must also have at least an after peak bulkhead enclosing the stern tubes and propeller shafts, and bulkheads at each end of the machinery space. The after peak bulkhead may terminate at the first deck above the load waterline provided that this deck is made watertight to the stern or to a watertight transom floor. Additional bulkheads ensure adequate transverse strength. Their total number, which depends on ship length, is laid down in Classification Society rules [4.6]. Fewer bulkheads may be fitted in ships which do not carry more than 12 passengers, if transverse strength is maintained in other ways, e.g. deep web frames. Bulkheads in passenger ships are governed by subdivision requirements described elsewhere (for British ships) [4.4, 4.7] and the bulkheads of Type A and B — ships must meet the relevant floodability requirements.

Transverse bulkheads must be watertight and must extend from the bottom of the ship to the freeboard deck. Thus if vehicles are carried below the freeboard deck and are required to pass through openings in the bulkheads, watertight doors are necessary. But where the main vehicle deck is also the freeboard deck, it may be completely clear of transverse bulkheads, other than the collsion bulkhead, to allow the unrestricted passage of vehicles.

Although all such requirements are intended to promote the safety of a ship, its crew and passengers, they also help to safeguard the cargo. Section

10.1.1 deals with the shipowner's obligations with respect to cargo.

Many of the current regulations were framed before Ro-Ro ships became popular, and so may not always take into account their special nature. IMCO is at present considering proposals for new requirements applicable specifically to them; Section 11.2.5 describes some possible implications.

4.4 Operational requirements

Unlike some of the earlier arrangements described in Chapter 1, modern cargo access equipment requires very little labour for its operation. Fig. 4.1 illustrates the large number of men required to open and close traditional hatches and the extremely dangerous areas in which they had to work.

Mechanical hatch covers are generally designed to open or close at a linear

Fig. 4.1. Closing a traditional hatch. In this ship the boards span only one beam space, although double length boards were common. Notice the man in the centre standing on a beam.

speed of about 10 metres per minute. Thus most hatchways can be covered or uncovered within one or two minutes, although final securing may take longer depending on cleating (securing) arrangements. In the past, steel covers were usually secured by screw cleats, but around 1960 the quick-acting cleat was introduced and later in the 1960s, the hydraulic automatic cleat became available, as described in Section 5.2.5.

When working cargo it is sometimes necessary to close hatch covers rapidly, perhaps because of the onset of a sudden rainstorm. If wires must be rigged for this, cargo may be damaged by the time the hatches are closed. Hatch-tents and covers that can be partially opened offer practical alternatives for preventing the entry of rain-water.

The shipowner is faced with a wide choice of access equipment and must take into account the type and size of his ship, its service and the size of its crew when selecting the equipment and deck layout that will best suit his particular operation. The selection of equipment is dealt with in greater detail in Chapters 7 and 8.

4.5 General considerations for access equipment

4.5.1 Coaming height

Minimum coaming heights, derived from the 1966 LLC, are laid down by Classification Societies. Heights are measured above the upper surface of the deck, and any sheathing that may be fitted, and for hatchways closed by portable covers secured weathertight by tarpaulins and battening devices, they must not be less than:

600 mm (23·5 in) for Position 1;
450 mm (17·5 in) for Position 2.

Coamings of hatchways closed by steel covers fitted with direct securing arrangements are usually as indicated above, taking into account any sheer or camber when assessing minimum height. They may, however, be lower, or even omitted entirely, if the safety of the ship is not impaired by so doing, and provided the Administration of the country concerned consents.

The scantlings and securing arrangements of flush hatch covers or those having less than standard height coamings are treated as special cases. Such arrangements have been approved in the past. Dock safety regulations generally require a minimum coaming height of about 760 mm (2 ft 6 in), otherwise additional fencing must be fitted to prevent personnel falling through the hatchway.

It is rare for a ship to have coamings of lower height than those stipulated, unless they are completely flush. Flush weather decks are required in a variety of circumstances. For instance, vehicles, wheeled cargo, or containers must sometimes be stowed over the full deck area including the hatch covers, and clear decks for recreation are desirable in passenger ships. Satisfying the floodability requirements for B−100 and B−60 ships may result in coamings being increased in height on such ships to meet load line and subdivision regulations.

In practice one of the principal determinants of coaming height is the operation of the hatch cover. Fig. 4.2 shows how it may be necessary to have the actuating mechanism below the cover, as in the case of hydraulic folding system. Other types that are affected in this way are roll stowing covers where the drum height must be sufficient to allow the stowed covers to fit between the drum axis and the deck (see Section 5.5) and single pull covers whose panels stow in an upright position standing clear of the deck as shown in Fig. 4.2.

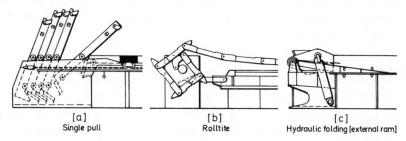

[a]
Single pull

[b]
Rolltite

[c]
Hydraulic folding [external ram]

Fig. 4.2. Cover stowing arrangements often determine coaming height. In (*a*) the coaming must be high enough for the single pull panels to tip, while the coaming for the roll-stowing cover (*b*) must be of comparable height to the stowage drum. The coaming for the folding cover (*c*) must be high enough to accommodate the external ram.

The coaming may also be increased in height to obtain additional cargo capacity. As a result of these considerations, the majority of ships have coamings that are considerably higher than the statutory minima, usually in the range 1·0–1·8 m. It is useful if crew and stevedores can see readily into the holds during port operations. Thus if the coaming is higher than about 1·4 m, a step or narrow platform should be fitted at a suitable height. While the top of the coaming is nearly always at the same level on all its sides, it need not be parallel with the deck. Neither is it essential that the coaming plates be vertical, if by sloping them inwards, a larger opening at deck level can be obtained.

Another factor influencing coaming height is the nature of particular deck cargoes. Packaged lumber, for example, is usually stowed abreast hatchways until it reaches the height of the hatch cover top, when it is distributed over the whole width of the ship. Lumber is usually banded into packages, each 660 mm high, which are stowed with a 25 mm batten between them.

Coamings should therefore be designed with a height from deck to hatch cover top which is a multiple of 685 mm (27 in). In heavy lift cargo liners, the coaming height and bulwark height are often the same so that awkward loads like barges may be easily supported across the entire ship.

In some ships the internal volume bounded by the coamings needs to be a greater proportion of the total hold volume than the minimum 600 mm coaming height allows. In the past certain ships engaged in the grain trade were required to have coamings enclosing 4% of the total hold volume to allow for settling, but this requirement has now been superseded by the provisions of the Safety of Life at Sea Convention 1974 (SOLAS 74) [4.7]. These new provisions do not stipulate a minimum or maximum coaming volume, although it may be necessary to arrange a coaming height in excess of 600 mm in a particular ship so as to meet the stability requirements for the carriage of grain. These assume that the cargo shifts into void spaces below side decks, so producing a heeling moment which depends on the resulting 'free surface' of the grain.

4.5.2 Cover stowage

Where a designer is attempting to obtain the largest possible hatchway openings, the question of hatch cover stowage is especially important. The width of the stowage space is fixed by the hatchway width if the covers are stowed at the ends of the hatchway, but the height and length of the space can be varied. If the hatchways are to be the longest possible in a given deck length, the length of the stowage space must be kept to a minimum; this can be accomplished in various ways depending on the type of hatch cover employed. No special stowage space is required for simple pontoon covers since these are usually stowed on adjacent hatch covers or on the quayside when the hatch is open. Alternatively side rolling covers may be used if the hatchway is not too wide in relation to the ship's breadth.

The ways in which the stowage space for various covers may be determined is discussed in Chapter 5, and the selection of hatch covers for specific purposes is discussed in Chapters 7 and 8.

4.5.3 Deck openings

As deck openings become larger, the problem of ensuring adequate hull strength becomes more complex. The necessary longitudinal strength of the hull girder can be readily achieved, even for ships with hatch widths 80% of the breadth of the ship, by the use of high tensile steels. But the provision of adequate torsional strength may require very detailed design and stress analysis, since the torsional deflection of a ship with large hatchways gives rise to high stress concentrations at the corners of the openings and to deformation of the hatchways.

To avoid high stress concentrations in deck plating, the corners of deck openings should be elliptical or parabolic. As the corners of the hatch covers are usually rectangular, provision has to be made to accommodate the hatch coaming. This can be done in two ways: either a square coaming can be built with the rounded deck plating protruding into the coaming space, as illustrated in Fig. 4.3 or the coaming can follow the shape of the deck opening with a filling-in plate welded to its top. In both cases the clear opening for rectangular cargoes like containers will be appreciably smaller than the maximum dimensions to the inside of the coamings.

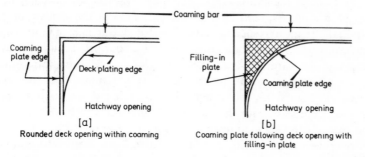

Fig. 4.3. Deck openings must have rounded corners to avoid stress concentrations. Two methods of matching a rectangular coaming to a rounded hatchway opening are shown. In (*a*) the rounded deck plating protrudes into the space bounded by the coaming, while in (*b*) the coaming plating is curved to follow the deck edge and a filling piece is welded into the coaming bar to provide a rectangular landing for the hatch cover.

4.5.4 Drainage

As mentioned earlier, an important function of access equipment is to prevent water entering the ship through the openings in its hull. If the hull is flexing in heavy weather, it is almost inevitable that some water will penetrate the seals of closing devices, especially if they are worn. Thus there must be a second line of defence, such as drains, for removing any water before it can damage the cargo.

Drainage facilities must be built into all items of access equipment, as illustrated in Fig. 5.12, for example. Here water which seeps past the peripheral seal of a hatch cover runs along a channel and is discharged onto the weather deck through a hole in the coaming. Where the drain is below the freeboard deck, as in flush weather deck covers, it must be connected to the bilge, or overboard via a scupper and non-return valve.

It is also necessary to provide drainage for the vehicle deck in Ro-Ro vessels. Any scupper which drains a space within an intact superstructure on the freeboard deck (usually the main vehicle deck) is led overboard through a pipe fitted with a screw-down non-return valve having open/shut indicators and capable of being operated from accessible positions above the freeboard deck. Alternatively the scuppers may be led down to the bilges or to a drain tank. In an enclosed tween-deck space with a continuous centre-line casing, additional scuppers must be installed adjacent to the casing on the main vehicle deck. Where the inboard end of a deck scupper would be below the load waterline at an angle of heel of less than 15°, it should be led to a separate drain tank, which may be pumped overboard.

4.5.5 Lashing of cargo

While there are no regulations regarding the strength of sea fastenings or the number of lashings required for a given item of cargo, the prudent operator gives special attention to such matters. The British Carriage of Goods by Sea Act and equivalent legislation in other maritime countries require the carrier to 'properly and carefully load, handle, stow, carry, keep, care for and discharge the goods carried'. Implicit in these requirements is the need to secure cargo against all movement. The securing of cargo has become increasingly important with the advent of container and Ro-Ro ships, and the practice of carrying containers or wheeled cargo on deck.

Modern practice for stowing containers on deck (as opposed to within cells in the hull) is to provide stacking points on the deck or on hatch covers which mate with the castings at each corner of the container via a portable

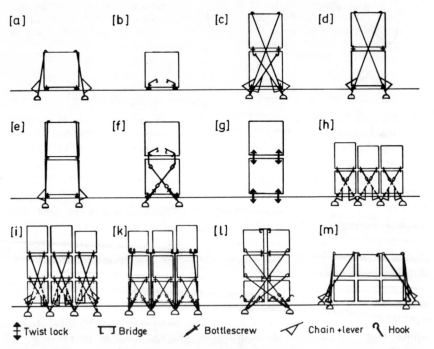

Fig. 4.4. Typical container lashing arrangements.

connection. Twistlocks, which are quick-acting connections between container corner castings, are then normally used to secure each container stack, together with bridges to hold adjacent stacks together and hooks, rods and bottle screws to lash the containers down. There is no mandatory arrangement of container lashings and different companies have developed their own individual patterns which vary according to the height and width of their container stows. Typical lashing patterns are shown in Fig. 4.4; the top tiers usually consist of either very light, or empty containers. Provision has to be made for supporting containers on the hatch covers. The lashing points are usually on the hatch cover but can be on the deck, depending on the number of containers, hatch size and ship layout.

In horizontal loading ships, the problem is not necessarily more complex but it is certainly more extensive, as every item of cargo, including those stowed on ramps, movable decks and other access equipment, should be lashed down or secured in some way. The most common lashing point is the 'elephant foot' (or clover-leaf) type (illustrated in Fig. 4.5) which can either

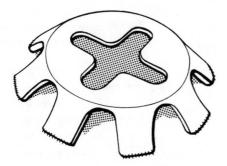

Fig. 4.5. An 'elephant's foot' or cloverleaf lashing point, used in large numbers in Ro-Ro vessels for securing wheeled cargo. They may also be recessed flush into the deck.

be recessed into the deck to allow the unhindered passage of vehicles, or be welded directly onto the deck plating. Lashing chains are then used to secure the cargo to the lashing point. Lashing points on the deckhead are often useful in Ro-Ros.

In any Ro-Ro vessel, there should be ample lashing points, adequately spaced to allow personnel to inspect and secure individual items of cargo during the voyage. It is common for suppliers to design lashing appliances to meet an operator's specific requirements, although standard equipment is now available for most purposes. Forces imposed on lashings by roll motions can be very large, as dynamic effects are included. Such forces can be appreciably reduced by roll stabilization systems, such as active fins, or special water-filled tanks, which are commonly fitted to Ro-Ro ships.

4.5.6 Maintainability

Poor maintenance has been found to be responsible for many instances where sea water has entered cargo spaces [4.9]. There are numerous factors which hinder the maintenance of cargo access equipment, not least of which is that the covers, doors etc. must usually remain closed at sea, preventing maintenance other than routine painting and greasing.

In port, it may also be difficult to do much maintenance because of the danger to personnel from moving cargo. Moreover, access to components in their stowed position may be restricted. This means that major maintenance work, such as the replacement of rubber gaskets or components of the hydraulic system, must usually wait for dry docking or survey periods when the ship is out of service for several days.

Modern access equipment has to be robust enough to operate satisfactorily with infrequent maintenance, but manufacturers try to design their products so that items like grease points are accessible in order to encourage regular attention from the ship's crew. Maintenance of cargo access equipment is discussed in more detail in Chapter 10.

4.6 Fabrication and installation

Until comparatively recently, it was usual for steel hatch covers to be made either by a shipbuilder under licence or by a sub-contractor to the hatch cover designer. This could lead to variations in the quality of the finished product. As hatch covers became larger and more complicated, and their variety increased, it became more common for specialist manufacturers to take direct responsibility for their fabrication as well as their design.

Manufacturer and shipbuilder must both work to dimensional tolerances which ensure that the covers can operate without major adjustments to the coamings. Final welding of coaming parts is usually deferred until the hatch covers are placed on board, to provide the basis for accurate alignment. To assist quality control, manufacturers have offered 'package deals' to supply hatch covers complete with coamings, and in some cases part of the surrounding deck plating too. These complete units are put aboard the ship in one lift and welded into place. This method has the advantage of ensuring that each cover matches its coaming perfectly. It also allows the manufacturer to test his product prior to installation and reduces the time required for fitting in the ship. Fig. 4.6 shows such a unit being lifted onto a ship. The final installation is carried out by shipyard personnel under the supervision of the hatch cover manufacturer.

The practice of building access equipment at one location and installing it in a ship at another has given rise to extensive transportation problems. With hatch covers measuring as much as 27 m × 20 m and weighing over

Fig. 4.6. A complete coaming and pontoon cover, supplied as a package deal, being installed aboard a new ship.

100 tonnes becoming increasingly common, it has become necessary to transport individual sections measuring say 27 m × 5 m or, in the case of complete coamings, much larger pieces. The usual practice is to fabricate the cover in sections which comply with the road haulage regulations of the particular country concerned, and then to assemble them in the shipyard.

Complete coamings are so large that they must be despatched by sea – by ship or barge. One hatch cover manufacturer has a special ship, classified by Lloyd's Register as a 'hatch cover carrier'. Offshore supply vessels are also used, as they have large open decks which are well suited to this purpose. They are also readily available for hire.

Cargo access equipment for Ro-Ro vessels is usually transported in a similar manner to hatch covers.

4.7 Safety

Hatches have always been a potential hazard to people on board ships and there have been serious accidents. For instance, a large side rolling cover, being opened by a single operator without assistance (even from a well-sited control position), could result in injury to someone standing on the far side of the hatchway, out of the operator's field of view. For this reason the

Authorities in many countries require that operators must check that the area in the vicinity of hatch covers is completely clear of all personnel before commencing opening or closing.

Unfortunately, in ships with large hatchways and high coamings, it is still possible for someone to walk into the danger zone without being noticed after the operator has checked that the area is clear. Such accidents are particularly likely on the weather deck where stevedores and personnel unfamiliar with the layout of the ship are likely to congregate at the start of a day's work. Thus many hatch covers now have warning alarms and lights to indicate that they are in motion. Alarms are also fitted to bow and stern ramps, doors and platforms in Ro-Ro ships to warn people who might be caught unawares.

An equally important safety consideration is hatchway fencing. In general, any area in which a person has to work, from which he could fall a distance of more than about 1·5 m, must be fenced. Tween-deck hatchways, flush weather deck hatchways, the tops of open, side rolling hatch covers in bulk carriers which must be cleared of grab spillage, and paired hatch covers on which cargo can be secured with one of the pair open, all require fencing. Practically all hatchway fencing has to be portable so as not to obstruct loading operations. Thus most fencing consists of posts about 1 m high which fit into sockets with wire or rope strung between them. This is not very satisfactory as it must be erected at every port and its components are easily lost.

Often fences are incorrectly set up. There should be two strands of wire, one at the top of each post and the other at mid-height, but the lower strand is often omitted, making it very easy for someone to slip under the top strand. A cheap and simple fencing that folds flush into a recess in the hatch cover top, and which could be lifted into position when required, would overcome some of these difficulties while forming a safe and readily available barrier.

Worldwide, there is an increasing tendency for more stringent safety precautions for cargo operations. Cargo access equipment must comply with any such requirements that may be appropriate.

References

[4.1] International Conference on Load Lines 1966. Inter-Governmental Maritime Consultative Organisation (IMCO) Publications, London.

[4.2] Merchant Shipping (Load Line) Rules 1968. Statutory Instrument 1053, H.M.S.O. 1968.

[4.3] Murray-Smith, D. R. The 1966 International Conference on Load Lines. *Trans Royal Institution of Naval Architects*, p. 3, 1969.

[4.4] Survey of Passenger Ships: Instructions to Surveyors. Department of Trade, H.M.S.O. 1961.

[4.5] The International Convention on the Tonnage Measurement of Ships 1969. IMCO Publications, London.

[4.6] Rules and Regulations for the Classification of Ships. Lloyd's Register of Shipping, London, 1978.

[4.7] International Conference on Safety of Life at Sea 1974. IMCO Publications, London.

[4.8] The Merchant Shipping (Tonnage) Regulations 1967. Statutory Instrument 172, H.M.S.O. 1967.

[4.9] Turnbull, D. E. Maintenance, Overhaul and Repairs to Hatch Covers. Joint Conference on Hatch Covers, Royal Institution of Naval Architects, Institute of Marine Engineers and Nautical Institute, January 1977.

CHAPTER FIVE

Access Equipment in Vertical Loading Ships

Summary

This chapter describes the principal types of hatch cover currently used in ships, together with their major characteristics. It includes single pull, folding, direct pull, roll stowing, rolling, lift and roll, sliding and pontoon covers.

5.1 Types of hatch covers

Many types of hatch cover are now available and the principal ones are listed in Table 5.1 where they are grouped according to their mode of operation.

Table 5.1. Operating mode of hatch cover types

Rolling and tipping	Single pull
Lifting	Pontoon
Rolling	End rolling
	Side rolling
	Lift and roll (piggy-back)
	Telescopic*
Roll stowing	Roll stowing (rolltite)
	Flexible rolling*
Folding	Hydraulic folding
	Wire-operated folding
	Direct pull
Sliding/nesting	Tween-deck sliding

*mainly fitted to inland waterway craft or barges.

Table 5.2 summarizes the most important characteristics of each major type.

Table 5.2. Characteristics of principal hatch cover types

Section	Cover type	Usual ship types	Decks applicable	Approximate % of covers fitted to recent new ships	Guide to minimum coaming height	Drive system	Cleating system
5.2	Single pull	(D)	Weather	41	Depends on section length but more than (A)	Electric, hydraulic or ship's cargo gear	(E)
5.3	Hydraulic folding	(D)	Weather/Tween	25	(A), (B) or (C)	Hydraulic	(E)
5.3	Wire-operated folding	(D)	Weather/Tween	10	(A), (B) or (C)	Winch	Screw or quick-acting
5.4	Direct pull	All ships with cargo gear	Weather	3	(A) or (B)	Ship crane or derrick	Automatic
5.5	Roll stowing	(D)	Weather	2	Depends on drum diameter but usually more than (A) or (B)	Electric or hydraulic	Automatic
5.6	Side and end rolling	All, but mainly large bulkers and OBOs.	Weather	4	(A) or (B)	Electric or hydraulic	(E)
5.7	Lift and roll	All, but mainly bulkers	Weather	2	(A) or (B)	Electric or hydraulic	(E)
5.8	Tween-deck sliding	Multi-deck cargo ships	Tween and Car decks	2	(C)	Electric	'Token'
5.9	Pontoon (F)	Container ships Multi-deck cargo ships	Weather/Tween	11	(A), (B) or (C)	Ship or shore crane	(E)

(A) Minimum coaming height allowed by 1966 Load Line Convention.
(B) (A) increased as necessary for fencing.
(C) Flush with deck provided drainage is satisfactory.
(D) All except combination carriers.
(E) Screw, quick-acting or automatic cleats can be fitted.
(F) Without tarpaulins on weather deck.

5.1.1 Definitions used in the text

Hatchway length. The fore and aft dimension of the clear opening ('daylight' opening).

Hatchway width. The athwartship dimension of the clear opening.

Hatch length. The fore and aft dimension of the hatch cover top plate.

Hatch width. The athwartships dimension of the hatch cover top plate.

Panel length. The fore and aft dimension of one hatch cover panel.

Panel width. The athwartships dimension of one hatch cover panel.

Panel depth. The depth of one hatch cover panel at centre.

Coaming height. The minimum vertical distance between the deck and the top of the coaming plate.

See Fig. 7.5 for typical dimensions

5.2 Single pull hatch covers

5.2.1 Description

The classic modern hatch cover is the 'single pull' which remains the most common of all the various forms now in service and may rightly be described as the natural successor to traditional beams and boards. This cover derives its name from its immediate predecessor, the 'multi-pull' cover, which consisted of a series of individual panels similar to those of the single pull,

Fig. 5.1. A single pull cover in the process of being closed. The panel lying flat on the coaming is pulling the next panel out of stowage.

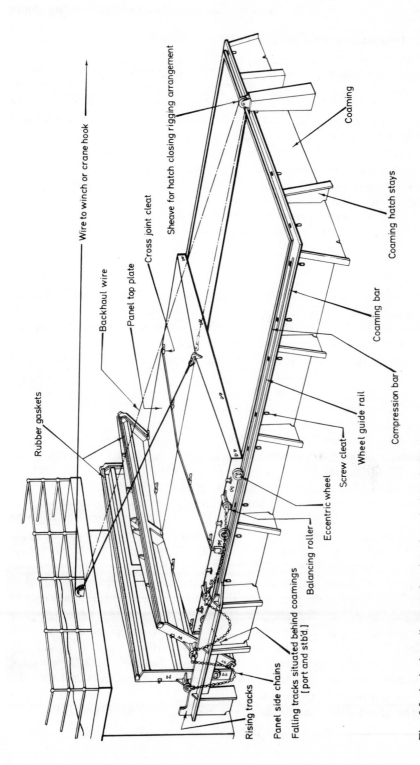

Wire to winch or crane hook

Sheave for hatch closing rigging arrangement

Cross joint cleat

Panel top plate

Backhaul wire

Rubber gaskets

Rising tracks

Panel side chains

Falling tracks situated behind coamings
[port and stbd.]

Balancing roller

Eccentric wheel

Screw cleat

Wheel guide rail

Compression bar

Coaming bar

Coaming hatch stays

Coaming

Fig. 5.2. A single pull cover showing the principal fittings.

but unconnected. Each panel had to be rigged before being pulled one at a time into stowage.

Fig. 5.1 illustrates the usual single pull arrangement. The complete cover consists of a number of narrow panels which span the hatchway and are linked together by chains. In the closed position, the panel sides sit firmly on a horizontal steel bar, attached all round the top edge of the coaming, which takes the weight of the cover. Just inside the side plates is a rubber gasket attached to the cover, which rests on a steel compression bar forming a weathertight seal. This is described further in Section 5.2.6. Extending from the side coamings at the ends of the hatchway where the covers are stowed are steel rails which allow the individual hatch panels to be transferred to their stowage location when the hatch is opened.

Although single pull covers rarely exceed 16 m in width, larger sizes can be manufactured.

5.2.2 Operation

To open a single pull cover, the securing cleats are first freed and each panel is raised onto its wheels by portable jacks and its eccentric wheels rotated through 180°. Alternatively, fixed hydraulic lifting gear may be used. In this condition the entire cover is free to move in a fore and aft direction with its wheels rolling between guides on the top of the coaming. It is set in motion by means of a wire (sometimes called a bull-wire) led from a winch and attached to the centre of the furthest edge of the leading panel, i.e. the panel which goes into stowage last and comes out of stowage first. As the wire is tightened, the panels are each pushed beyond the end coaming. On reaching the end of the hatchway, the weight of each panel is transferred from its wheels to balancing rollers (see Fig. 5.3) situated near its midlength which engage with the rail extensions of the side coamings. The centre of gravity of each panel is slightly towards the stowage end of the panel from the rollers so that once the panel is supported on them, it tips into the vertical position. It is then pushed further towards the end of the stowage space by the next panel to arrive at the hatchway end. When the hatch is completely open, each panel stands vertically in the stowage space, and all are kept in place with retaining chains, or hooks.

To close a single pull cover, the securing chains are removed and a back-haul wire, which during operation should always be attached to the cover in the same place as for opening, runs through a sheave so that when tightened the panels are pulled back over the hatchway. As the first panel leaves its stowage position, it tips through 90° about its rollers to land horizontally with its wheels resting on the coaming. It is then pulled further over the hatchway and the chain linking it to the next panel becomes taut with the result that it too is set in motion.

On reaching the end of the stowage space, the second panel tips to lie

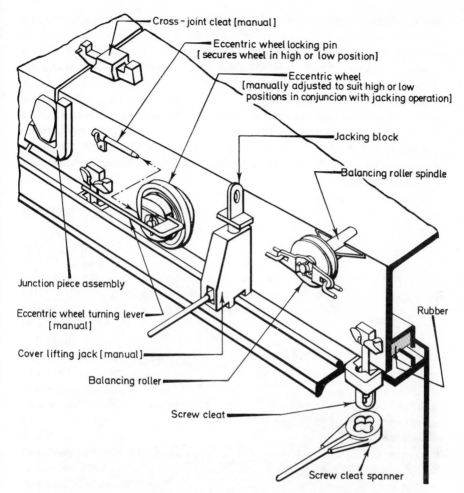

Cross-joint cleat [manual]

Eccentric wheel locking pin
[secures wheel in high or low position]

Eccentric wheel
[manually adjusted to suit high or low
positions in conjuncion with jacking operation]

Jacking block

Balancing roller spindle

Junction piece assembly

Eccentric wheel turning lever
[manual]

Cover lifting jack [manual]

Balancing roller

Screw cleat

Rubber

Screw cleat spanner

Fig. 5.3. Detail of a single pull cover side plate. The lifting jack is portable and is removed when not in use.

horizontally on the coaming behind the first and together they progress across the hatchway. The process is repeated with the third and subsequent panels, ending only when all are lying flat on the coaming and the hatchway is completely covered. During the tipping process, the leading edge of the tipping panel engages in the trailing edge of the panel ahead of it, locking them together at the cross-joint.

5.2.3 Construction

Each panel of a single pull cover is usually constructed as a simple mild steel plate, stiffened by a series of beams spanning its width with a vertical skirt around its perimeter as shown in Fig. 5.4 and Fig. 10.3. Its upper surface may be flat or peaked with a central ridge. Both forms can be designed to support additional loads imposed by deck cargo (for which flat-topped panels are more suitable). Peak-topped panels are slightly lighter. This is because the beam height at the centre of all types of hatch cover of similar construction is typically 4% of the hatch cover span, while at the sides it is 2% for the peaked cover, 4% for the flat-topped version. This reduction in depth is made possible by the reduced bending stresses towards the end of the beam. Usually the panels are not plated on the underneath, although one manufacturer promotes double-skinned covers (see Sections 9.1.3 and 9.1.4).

Fig. 5.4. Single pull covers fitted in a bulk carrier. The construction of a panel with its transverse beams and circular ventilators is clearly visible. The sheave post for opening and closing wires can be seen between the stowed covers. The ship is discharging coal.

5.2.4 Stowage

Stowage space for individual hatch panels must be provided at one end of the hatchway, or at both ends of long hatchways where split covers are fitted. The minimum dimensions of the stowage space depend mainly on the length and depth of the panels, and their total number.

5.2.4.1 *Length of stowage space.*

Since the panels usually stow vertically, the fore and aft length of the stowage space is given approximately by

$$\text{Stowage length} = (0.05 \times S \times N + 0.37 \times L)$$

where N is the number of panels stowing at the particular end, S is the athwartships span of the cover (generally hatch width) and L is the panel length, in metres. N is chosen to be in the range 2–11 but is usually 5 or 6 and is such that L, which is the hatch cover length divided by N, is in the range $0.2\,S$ to $0.3\,S$. A further 0.5 m should be allowed for cross-deck access for personnel if required.

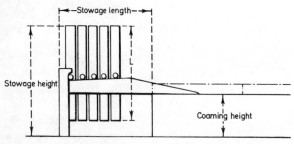

Fig. 5.5. Single pull cover stowage space. See also Fig. 7.5.

5.2.4.2 *Height of stowage space*

Single pull covers are often stowed under raised winch platforms where headroom is restricted. Minimum headroom above the deck is given approximately by panel length (L) + clearance margin. The clearance is usually in the range 300–500 mm.

Coaming height is related to panel length; longer vertical-stowing panels can only be used if the coaming height is raised. In principle coaming height is approximately given by $L/2$. This is usually greater than the minimum coaming height of 450 mm or 600 mm depending on hatch position, laid down by the 1966 Load Line Convention (see Section 4.5.1). A high coaming has the advantage of permitting fewer panels, thus reducing the number of cross-joints to be maintained.

The ideal stowage space can be achieved by judicious matching of panel length and depth and coaming height, and can be designed by a hatch cover manufacturer to meet an owner's specific requirements.

In some ships it may be necessary for coaming height to be low, perhaps little more than standard height, and the usual arrangement of stowage would require very short panels and hence a long stowage space and many cross-joints. Longer panels can be used in such circumstances provided that their rollers are located a distance approximately equal to the coaming height from their lower edges when stowed. A consequence of this is that they are unbalanced and will not tip when opening without the application of an external force which can be applied by rods substituted for the chain linkages between adjacent panels. Fig. 5.6 shows a rod-operated design.

Fig. 5.6. Rod-operated single pull cover under test prior to installation in a ship.

5.2.4.3 *Width of stowage space*
Hatch panels are stowed within the limits of the side coamings and supports, typically 0·5–0·75 m wider than the hatchway on each side. Hatchway width is set by overall considerations of cargo access (see Chapters 7 and 8).

5.2.4.4 *Variants arising from stowage considerations*
Several modifications to the standard single pull cover have been introduced to reduce the stowage space required at the hatch ends. In one arrangement, panels are constructed as inverted dishes or pans (see Fig. 5.7). The shape

contributes to the strength of the covers and allows them to nest together when standing vertically, thereby reducing the stowage length necessary.

Fig. 5.7. Single pull cover with close-nesting pan-type panels. See also Fig. 2.3.

In another variant (M type), used in ships where both the horizontal and vertical stowage space is limited, the athwartships edge nearest the stowage end of the panel is constructed with a deeper beam which results in nesting and a reduction in stowage length. These panels stow in a slightly inclined position as indicated in Fig. 5.8.

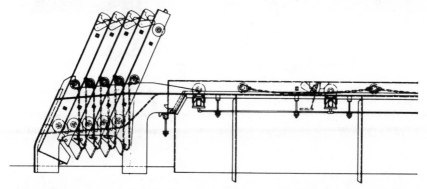

Fig. 5.8. 'M' type single pull cover in the stowed position with each panel inclined. A deeper beam at the trailing edge permits a shallow construction over the rest of the panel, with the overall effect of reducing stowage space requirements.

5.2.5 Cleating arrangements

Before any ship can proceed to sea, its hatch covers must be secured in the closed position to ensure that they remain weathertight. This can be accomplished in a variety of ways depending on the degree of automation required.

5.2.5.1 *Screw cleats*

One of the earliest methods of cleating was by means of a large number of closely spaced screw cleats (or dogs) for peripheral locking and wedge type cross-joint cleats, both of which are shown in Fig. 5.3. These cleats are manually operated, the peripheral ones by means of a special spanner and the cross-joint wedges by being knocked home with a mallet.

5.2.5.2 *Quick-acting cleats*

A faster means of securing panels to the coaming is provided by quick-acting

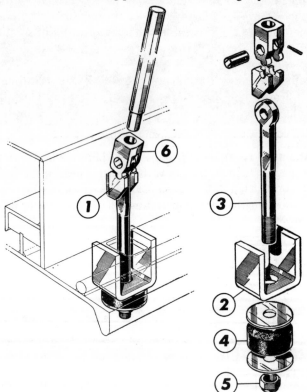

Fig. 5.9. A quick-acting cleat showing its various parts and operating lever. Key: (1) Snug welded to cover. (2) Crutch welded under coaming bar. (3) Threaded rod. (4) Neoprene washer. (5) Nut. (6) Hinged head. See also Fig. 5.1.

cleats (Fig. 5.9) which are operated manually by means of a special lever acting on an 'up and over' principle. The amount of compression provided by this arrangement is pre-set by means of neoprene washers so that all cleats are tightened down evenly and cannot be over-tightened. The cleats can be quickly disengaged from the cover snugs, and stowed in recesses in the coaming face bar. Classification Society experience shows that widely spaced cleats, up to 2 m apart can be satisfactory.

Tests to compare the operation of quick-acting and screw cleats have been performed. In one case, a hatch cover with 39 cleats was secured in 17 minutes using screw cleats, while only 6 minutes were required when quick-acting cleats were fitted.

5.2.5.3 *Automatic cleating*

Securing may be performed even more quickly by means of completely automatic peripheral and cross-joint cleating arrangements. Peripheral cleating can be operated by one man in under two minutes using either a motorized or manual hydraulic pump. Automatic cross-joint cleating is achieved by means of a torsion bar fitted between each pair of panels. Lever arms at the outer ends of the bar engage with the coaming as the cover closes, rotating the torsion bar and pressing evenly-spaced cleating pads mounted along it against lugs attached to the adjacent panel. This process is illustrated in Fig. 5.10.

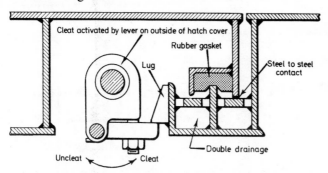

Fig. 5.10. Section through an automatic cross-joint cleating arrangement.

In another form of automatic cross-joint cleating, a transverse bar is drawn across the hatch by means of a manually operated threaded wheel.

Automatic peripheral cleating is provided by means of sliding bars fitted along the side and end coamings under the top rail. Hooks are attached to the bars at each cleating point where there is a slot in the coaming rail (see Fig. 5.11). The bars are operated by double-acting hydraulic cylinders, and slide horizontally to raise the hooks upwards and through the slots where they engage in the cleat lugs on the cover sides and ends. The action is reversed to uncleat.

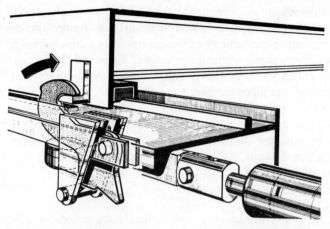

Fig. 5.11. Automatic peripheral cleat for a single pull cover.

5.2.6 Weathertightness

As discussed in Sections 4.1.5 and 10.1.1, weather deck hatch covers are required to be weathertight. This is accomplished by means of sealing gaskets and drainage channels (see Fig. 5.12).

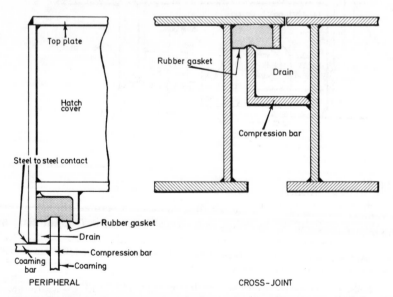

Fig. 5.12. Single drainage and seals for a single pull cover, cross-joint and peripheral.

5.2.6.1 *Single drainage*

The initial defence against sea water entry is the extremely small gap between the cover and coaming and between adjacent panels. Water penetrating these gaps is drained away along the drainage channels and out through the side of the covers onto the deck. Drainage channels are sealed by means of rubber gaskets and compression bars which provide the principal defence.

5.2.6.2 *Double drainage*

A further refinement of this system is the provision of a second drain channel which ensures that any sea water that gets past the seal is drained away via a non-return valve in a similar fashion to that in the first drain, as illustrated in Fig. 5.13.

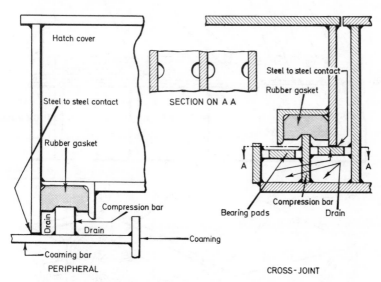

Fig. 5.13. Double drainage and seals for a single pull cover.

5.2.6.3 *Seals and compression bars*

The most common form is that shown in Figs. 5.11 and 5.12, but this can be refined by substituting a round section stainless steel compression bar which suffers less from corrosion, as illustrated in Fig. 9.5.

5.2.7 Drive systems

The system already described (Section 5.2.2) is known as a wire-operated drive, which is both the simplest and cheapest arrangement, but also the slowest in operation because of the time needed to rig the wires. All hatch cover drive systems are generally designed to operate with ship trims of up to 2° by the stern.

5.2.7.1 *Fixed chain drive*

Operating times can be reduced by self-powered systems in which a pair of electric motors in the leading cover panel (first out of stowage) drive sprockets on each side of the cover which are permanently engaged with a fixed chain. When in motion, the leading panel pushes or pulls the other panels across the hatchway. Power for the electric motors is supplied by a

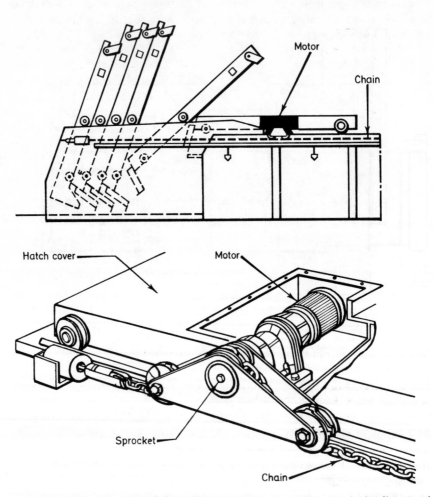

Fig. 5.14. Fixed chain drive for a single pull cover. A sprocket on the leading panel engages in a fixed chain resting along the coaming bar and is driven by a motor as shown to pull or push the other panels for closing or opening.

cable from the ship's supply. Alternatively the sprocket and chain can be replaced by a rack and pinion.

5.2.7.2 *Long chain drive*

A further self-powered alternative employs endless chains running externally along the full length of both side coamings, connected to the leading panel. The chains are led via sprockets to an external electric or hydraulic motor. Detachable towing connections allow a single drive unit to operate more than one set of covers.

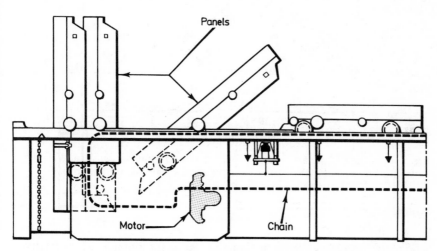

Fig. 5.15. Long chain drive. In this case the leading panel is fixed to a continuous chain powered by an external drive unit.

5.2.8 Weight

Manufacturers supply a figure for the weight of cargo access equipment when enquiries are being made, but it is useful for naval architects, shipbuilders and operators to be able to obtain a quick estimate of the weight of proposed hatch covers. Fig. 5.16 shows how the weight per square metre varies with the span for typical single pull covers ranging from about 0·14 to 0·24 tonnes. Heavy deck cargoes generally lead to increased weight, but this depends on hatch cover length, span and type, with covers for two tiers of containers weighing approximately 20–25% more. The additional structure necessary, such as stowage ramps, amounts to roughly an additional two tonnes per stowage end. Although coamings are usually supplied by the shipbuilder, an approximation to their weight in tonnes per metre run is given by 0·18 h where h is the coaming height in metres.

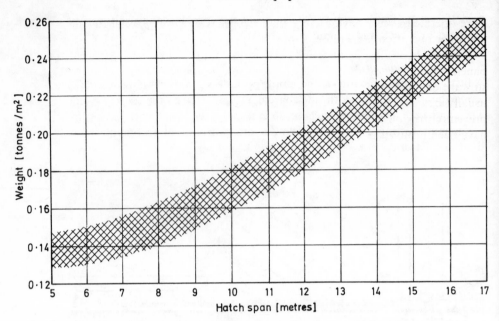

Fig. 5.16. Single pull panel weight per square metre as a function of span. Weight will vary with number and length of panels, generally within the range shown.

5.2.9 Applications

The main advantage of the single pull cover is its simplicity. It is relatively inexpensive and maintenance and repairs can be carried out fairly easily. This accounts for its popularity as shown in Table 5.2.

Single pull covers are also extremely adaptable, and have been installed on weather decks in large and small ships ranging from all sizes of general cargo and refrigerated vessels to bulk carriers. Their prinicipal drawbacks are their noisy operation and their need for stowage space at the hatchway ends. In their simplest form, cleating is slow.

5.3 Folding covers (hydraulic or wire-operated)

5.3.1 Description

Folding covers may be fitted at both weather deck and tween-deck hatchways. In its simplest form, this type of hatch cover consists of two flat-topped panels, similar in basic construction to those of the single pull system. Complex configurations may have three or more panels at each end of the hatchway, although installations with an uneven number of panels are rare.

Wire-operated covers having more than two panels require special rigging and their operation is therefore slow.

A typical hydraulically-operated cover is illustrated in Fig. 5.17. Adjacent panels are hinged together so that they can fold as shown. The panel at the stowage end is hinged to a plinth welded to the deck and hydraulic rams are usually arranged as illustrated. The ram rod is withdrawn into the cylinder to prevent corrosion when the hatch is closed at sea. Covers up to 26 m wide have been installed in ships.

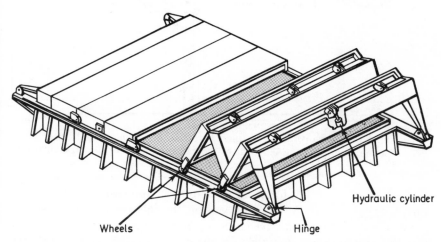

Fig. 5.17. A multi-panel end-folding hydraulic cover for weather deck use.

5.3.2 Operation

To open an hydraulic folding cover, the cleats are released and then the actuators are operated, causing the hinge point between the panels to rise. The leading edge of the leading panel is supported by wheels on both side coamings. The hydraulic ram extends until the cover is in its stowed position.

A multi-panel installation is opened in the same way, with the pair of panels at the stowage end always folding first, the leading panel last. This is accomplished by suitable programming of the hydraulic valve system, which is done by the manufacturer before the cover is installed.

A wire-operated cover is opened by means of a bull-wire led from a winch or crane, through a sheave at the stowage end of the hatchway as shown in Fig. 5.18. By pulling on the wire, the cover folds open in exactly the same way as its hydraulic counterpart.

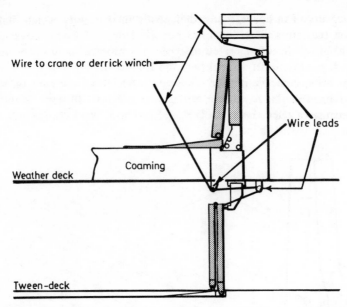

Fig. 5.18. Wire-operated end folding weather and tween-deck hatches in the open position showing the leads of the wires to cranes or derricks.

To close the covers, the foregoing operations are reversed. The appropriate valves are operated on the hydraulic version, the bull-wire released slowly on the wire-operated type, and the covers close under gravity. Fig. 5.19 shows that the panels do not stow tight up against each other. This is to ensure that the covers will readily start to roll into the closed position as the wire is released.

5.3.3 Construction

The construction of folding panels can be virtually the same as for the single pull system or alternatively a grillage arrangement may be adopted. They may be strengthened by deeper or heavier beams and plating according to the type of deck cargo they are required to support. Where side folding covers are fitted in more than one fore and aft section an intermediate cross-beam support is required, as indicated in Fig. 5.20.

5.3.4 Stowage

As for single pull covers, stowage space depends on the length, depth and width of the individual panels, and sufficient stowage space must be provided at hatchway ends to allow a completely clear opening. The coaming height is usually greater than that laid down in the 1966 Load Line Convention to allow sufficient room for the actuating ram on hydraulically operated covers. The space required depends on ram size which in turn depends on the size and weight of the cover. For flush-fitting covers, the coaming height is zero but for other types it is more usually in the range 1–2 metres as discussed in Section 4.5.1. Individual panels on each installation are not necessarily the same length but they must have the same depth and breadth. In particular, the first panel into stowage is usually shorter because of the hinged arms which must span the stowage area.

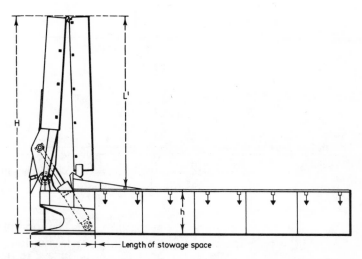

Fig. 5.19. Hydraulic end-folding weather deck cover showing stowage requirements. Notice that the panels are of different lengths because the fulcrum is some distance from the end coaming.

5.3.4.1 *Length of stowage space*

This is determined by the number of panels, their depth and the angle between them when stowed. Hydraulic actuators can be fitted to ensure that the panels fold tight against each other and these are discussed further in Section 5.3.7. Thus the minimum length of stowage space is given by the panel depth multiplied by the number of panels. For any given hatchway, the required length of stowage space is less for a folding cover than for an equivalent single pull cover.

5.3.4.2 *Height of stowage space*

The total headroom required is

$$H = L' + h$$

where h is the coaming height and L' is approximately $\dfrac{1\cdot3 \times \text{hatch length}}{\text{number of panels}}$

The stowage height is considerably greater than that required for single pull covers, but it can be reduced by installing a multi-panel cover although this necessitates an increase in stowage length.

5.3.4.3 *Width of stowage space*
As for single pull covers, the width of stowage space is the width of the hatch coaming plus supports.

Fig. 5.20. Side-folding, flush-fitting, hydraulically operated tween-deck covers. Covers can be arranged to form a vertical division when open.

5.3.4.4 *Variations arising from stowage considerations*

Tween-deck covers are sometimes arranged to fold towards the sides rather than towards the ends of the hatchway (Fig. 5.20), forming a vertical division if required. It is usually possible to design folding covers to suit the specific requirements of any particular ship.

5.3.5 Cleating

All the cleating arrangements discussed for the single pull system (Section 5.2.5) are applicable to folding covers situated on the weather deck. When used on the tween-deck as non-weathertight covers, the classification societies require the fitting of 'token' cleats at each end of the cover. At the stowage end, the hinges are acceptable as cleats, but the leading panel is fitted with one screw-down cleat at each side.

5.3.6 Weathertightness

Folding covers, like single pull covers, are kept weathertight by means of gaskets and drainage channels, but arranged as shown in Fig. 5.21.

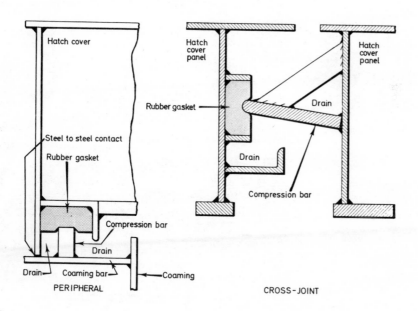

Fig. 5.21. Double drainage and seals for hydraulic folding covers.

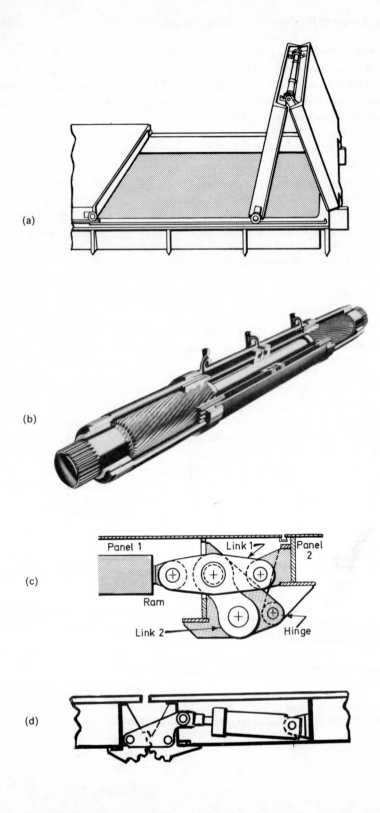

(a)

(b)

(c)

Panel 1

Link 1

Panel 2

Ram

Link 2

Hinge

(d)

5.3.7 Drive systems

5.3.7.1 *Hydraulic hinges*

In certain circumstances, it may be desirable to have a folding cover in which the panels stow tight against each other. Special hydraulic actuators are necessary in such circumstances. These can take various forms as illustrated in Fig. 5.22. Fig. 5.22a and b shows a self-energizing hinge which minimizes stowage space, and Figs. 5.22c and d show ram and link arrangements which require slightly more horizontal stowage space, because of the link geometry.

5.3.7.2 *Multi-panel, exterior ram folding covers*

For ships which carry cargoes that would be damaged by any leak of hydraulic oil (e.g. bulk grain), folding multi-panel covers operated by external hydraulic rams are available. These ensure that any oil leak, however unlikely, spills onto the deck and not the cargo. External rams are easily fitted to twin-panel covers (see Fig. 5.17) but for multi-panel covers, which are often required in large bulk carriers, a completely different arrangement is necessary as shown in Fig. 5.23. Here a four-panel arrangement is operated by four hydraulic rams and a series of robust levers. This cover can only be used on weather decks.

5.3.7.3 *Hydraulic oil supply*

Pressurized oil for hydraulic folding covers can be supplied from either a central power pack located in a deckhouse which is linked to all the covers in a hatchway, or it can be supplied from 'mini-packs' connected to each individual pair of covers. These are self-contained hydraulic pumps and oil reservoirs. In both cases, the oil is conveyed via steel and flexible rubber piping to the rams. The connections between the panels are illustrated in Fig. 5.24. Alternatively in multi-panel installations with internal rams, each ram can be supplied with oil from a mini-pack situated within the panel.

5.3.7.4 *Wire-operated*

These use a bull-wire, attached to a ship's crane or derrick which opens or closes the covers. It is usually a long and difficult process to rig bull-wires

---

Fig. 5.22. Hydraulic operating mechanisms for folding covers which do not employ external rams: (*a*) and (*b*) Navire Hydrautorque hinge. Hydraulic pressure introduced at the centre of the hinge forces the two pistons apart, so that the splined shafts are rotated through 180° to open (or close) the pair of covers. (*c*) Ram and link mechanism. Actuating the ram and connecting link (1) allows the panels to fold 180° about the hinge, with the link (2) controlling the movement. (*d*) Ram and cog mechanism, which can be used in insulated tween-deck covers.

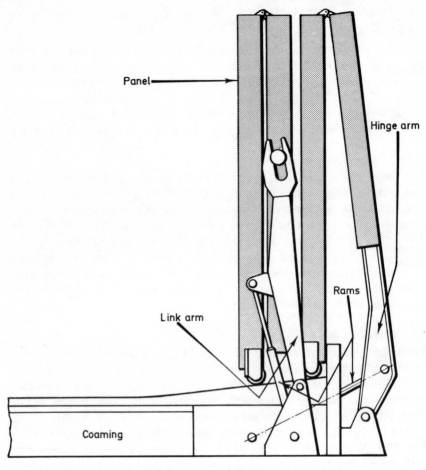

Fig. 5.23. Multi-panel external ram, hydraulic folding cover, of Kvaerner Multifold Crocodile type. This arrangement allows four panels to stow at one end of the hatch-way, operated only by external rams.

to tween-deck folding covers, particularly on ships with more than one tween-deck, unless special arrangements are made.

5.3.8 Weight

Folding covers are slightly heavier than single pull covers for the same hatch-way as shown in Fig. 7.5, but there is no additional weight for stowage rails. The weight of tween-deck covers depends on their wider variety of loadings. Covers with internal rams are heavier owing to their increased internal stiffening.

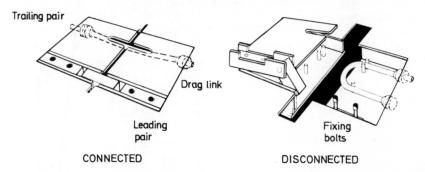

Fig. 5.24. Tween-deck hatches are often required to be only partially opened. The arrangement shown allows folding panels to be disconnected and isolated for this purpose.

5.3.9 Applications

Folding covers are becoming increasingly popular for both weather and tween-deck use. Side folding covers in the tween-deck can be designed to form grain-feeders or other divisions if desired, as shown in Fig. 5.25. End folding covers are sometimes used on the weather decks of bulk carriers as their horizontal stowage space requirement is small. Short single-panel covers folding through 180° are sometimes used on Great Lakes ships. Folding covers can easily be adapted for hydraulic automatic cleating. Their main advantages, however, are their easy and quick operation and the short horizontal stowage space which they need.

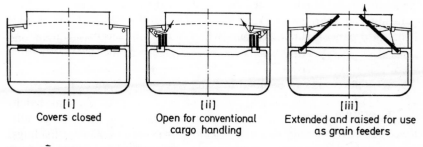

Fig. 5.25. Wire-operated side-folding covers which can be used as grain feeders when inclined as shown.

5.4 Direct pull covers

5.4.1 Description

Direct pull covers are a recent development, usually found in general cargo ships where multi-panel covers have to be operated by a wire, using ship's gear. Three-panel covers are usual and one is shown in Figs. 5.26 and 5.27. The panel at the stowage end (*1*) is fitted with a hinged central sheave (*s*) and connected to the ship's structure by means of two hinged stowing arms (*l*). The second panel (*2*) is hinged to the third (*3*) at its bottom edge (*hi*).

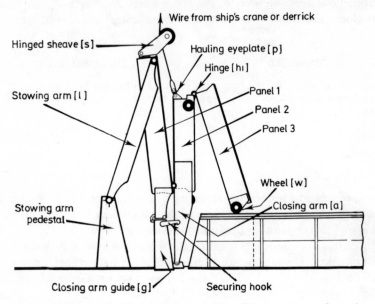

Fig. 5.26. Direct pull cover in the open position. The key letters refer to the text in Section 5.4.

The second panel carries a fixed closing arm (*a*) which, during the opening operation, engages a guide (*g*) and is deflected downwards. The trailing edge of the third panel is supported on a wheel (*w*) which runs on a track at the top of the coaming. All the above components, except the sheave, are fitted to both sides of the cover.

Direct pull covers can be fitted to hatchways up to 20 m or more wide.

5.4.2 Operation

Direct pull covers do not require uncleating before opening as their cleating system is fully automatic and is activated in the opening sequence. The bull-wire is rove through the sheave, attached to the hauling eyeplate (*p*) when the cover is installed, and is not normally removed. Its free end is attached to a derrick runner or crane fall and is pulled to open the cover. This operation raises the sheave (*s*). At the same time, the other two panels attached at the hinge (*hi*) move towards the stowage end, but remain flat. When the end of the closing arm (*a*) comes into contact with the guide (*g*), it is deflected downwards raising the hinge (*hi*). All three panels continue to move together until they reach their stowage position where they are locked in place by hooks as shown in Fig. 5.26 and the crane hook is released. To lower the panels, the crane hook is re-attached, the strain taken on the wire, the securing hooks released and the cover is lowered under gravity. On reaching their closed positions the panels are automatically cleated, ready for sea.

Fig. 5.27. Direct pull cover being operated. Note the construction of the panels and coaming.

5.4.3 Construction

Direct pull covers are of a normal stiffened-plate construction as shown in Fig. 5.27.

5.4.4 Stowage

Stowage space depends on the length, depth and width of the panels. The total space required depends more on the geometry of individual direct pull covers than is the case for single pull and folding covers, as the panels stow at different heights and with an angle between them. The gap at deck level so created may allow the unimpeded cross passage of personnel. A general formula is not applicable, but the length requirement is less and the height requirement is greater than for single pull stowage as shown in Fig. 7.5.

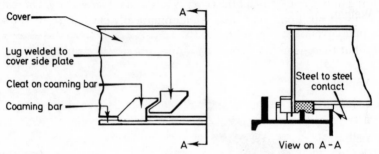

Fig. 5.28. Automatic peripheral cleating and sealing arrangement for a direct pull cover.

5.4.5 Cleating

The peripheral and cross-joint cleating of this type of cover is part of the closing process. Peripheral side cleating and end cleating is accomplished by means of lugs welded to the coaming and cover side plates as illustrated in Fig. 5.28.

The ends and cross-joints are secured by 'finger' cleats which interlock as the panels come together at the cross-joints. The cleats at the end of the cover, remote from stowage, are shown in Fig. 5.29.

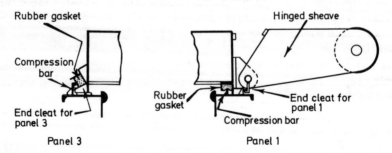

Fig. 5.29. Automatic end cleating and sealing arrangement for a direct pull cover.

5.4.6 Weathertightness

Weathertightness is achieved by means of gaskets and drainage channels similar to those on other covers. The leading panel gasket is external, unlike other cover types, and is shown in Fig. 5.29.

5.4.7 Drive system

The covers are opened and closed by means of a wire from a crane or derrick as already described.

5.4.8 Weight

Direct pull covers are slightly heavier than single pull covers as shown in Fig. 7.5 but there is no additional allowance for stowage rails.

5.4.9 Applications

These covers are used on the weather decks of general cargo ships and geared bulk carriers, although they could be fitted to other types of vessel with cargo gear. They are restricted to such ships because of the need for cranes or derricks to operate them. They are considerably cheaper than equivalent hydraulic folding systems. Their first cost is no greater than that of most single pull systems as indicated in Table 7.1 but, because of their automatic cleating, they are quicker and easier to operate. They cannot normally be used in situations where the cover must stow under a deckhouse overhang. In this case a single pull cover is probably the most suitable alternative.

5.5 Roll-stowing covers – 'Rolltite'

5.5.1 Description

Roll-stowing covers are often called 'Rolltite', the trade name of a type of hatch cover supplied by MacGregor, which has panels that roll on to a drum for stowage. The cover has been developed from a roll-stowing cover originally designed by Ermans. It consists of a number of panels spanning the hatchway, each pair of which is longer than the pair immediately before it, viewed from the stowage end. The leading panel is supported by a single wheel resting on a track at each side. The track is inclined towards the coaming top at the stowage end. Each panel has a horizontal and vertical section as illustrated in Fig. 5.30.

The pressure pad or roller of the vertical section rests on an inclined ramp situated on the inside of the coaming (see Fig. 5.31). A stowage drum is situated at one end of the hatchway. This extends over the full width of the hatchway and is connected to the cover by means of bridging arms at both sides. The drum is mounted on a structure situated some distance from the

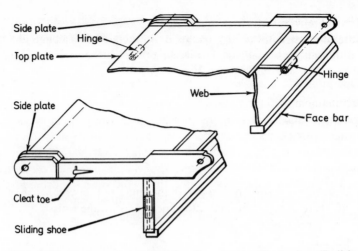

Fig. 5.30. Constructional details of a single roll-stowing (Rolltite) panel. Notice that practically all of the panel's strength is derived from a single vertical web.

Fig. 5.31. Open hatchway fitted with a roll-stowing cover. Note the ramp fitted along the side coaming. This raises the panels as they move towards the stowage drum.

end of the hatchway, which also contains the powering mechanism.

Rolltite covers are not usually installed at hatchways greater than 17 m in width, although greater spans are possible.

5.5.2 Operation

The Rolltite cover does not need uncleating or lifting before opening. The operator simply starts the driving motor and the cover is slowly wound onto the stowage drum. As the cover begins to move, the wheels at its leading end rise up the small ramp, the cleats unfasten and the vertical portions of each panel in contact with the ramp rise up so that the cover is lifted off its sealing gaskets. The panels are designed so that they wind onto the drum as shown in Fig. 5.32. Throughout the opening process, three panels are suspended between the stowage drum and the end coaming.

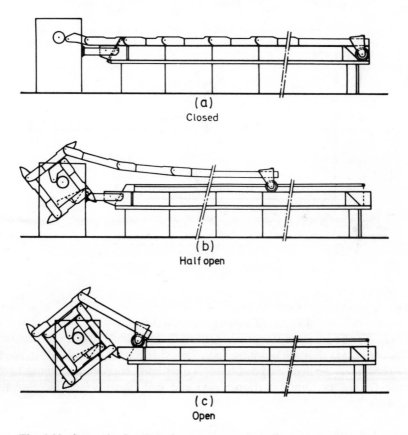

(a)
Closed

(b)
Half open

(c)
Open

Fig. 5.32. Stages in the operating sequence of a roll-stowing cover. From the fully closed position (*a*) the cover winds on to the stowage drum, one panel at a time, until it is fully open (*c*).

5.5.3 Construction

The construction of Rolltite panels is different from all of those already described. Each panel is constructed as a beam with a wide top flange, as shown in Fig. 5.30. The vertical web carries the two sliding shoes (bearing pads) or wheels which support the panel as it rises up the sloping ramp. It has to lie against the inner surface of the top plate of the succeeding panel which can therefore be only lightly stiffened. Thus, each panel derives its strength entirely from the top and side plates and the vertical stiffener.

5.5.4 Stowage

The stowage drum is located so that the stowed cover can rotate freely, clear of the coaming and the deck. Fig. 5.33 shows this, together with the stowage of an equivalent single pull cover for comparison.

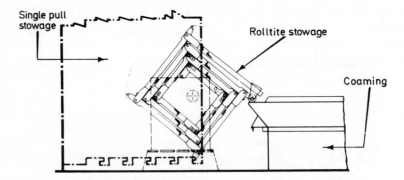

Fig. 5.33. Stowage space required for roll and stowing covers compared with that for an equivalent single pull cover. See also Fig. 7.5.

5.5.4.1 *Length of stowage space*
This is given approximately by
$$\text{Length of stowage space} = 0.60\sqrt{(l+3)}$$
where l is the hatchway length in metres. This formula applies to standard Rolltite covers that are not modified to provide for the extra loads imposed by containers or deck cargo. Covers which must support additional loads have thicker plating and deeper side and end plates which increase the stowage length.

5.5.4.2 *Height of stowage space*
From Fig. 5.33 it is clear that the height of the stowage space is for practical purposes the same as its length.

5.5.4.3 *Width of stowage space*

This is slightly more than the width of the coaming because the drum supports and driving mechanism are outboard of the drum itself. An additional margin of 1 m is usually sufficient.

5.5.5 Cleating

Rolltite covers are secured automatically during the closing process. Peripheral cleating is by means of lugs welded to the side plates of the panels which engage in shoes, attached by bolts and neoprene washers to the coaming as the panels move through the last 50 mm of their travel. See Fig. 5.34.

At the end remote from stowage, pads on the final panel slide under fingers welded to a bar attached to the coaming by means of bolts and neoprene washers. The purpose of the neoprene washers is to provide a uniform cleating pressure all round the cover and prevent overtightening of the cleats. At the stowage end, the cover is secured by the rotary movement of the panel as it closes. This allows the finger cleats to move into the holes in the panel. Fig. 5.35 shows the cleats at the other end. Cross-joint cleating is provided by the hinge pins between the panels.

Fig. 5.34. Automatic periphal cleating for a roll-stowing cover.

Fig. 5.35. Automatic end cleating for a roll-stowing cover.

5.5.6 Weathertightness

The peripheral sealing arrangement of the Rolltite cover uses a standard system of gaskets and compression bars as illustrated in Fig. 5.36 with the drainage channel doubling as the internal ramp. The cross-joint seals are unique, however. Adjacent panels are never separated from one another, and therefore the cross-joint seals are permanently made by means of sheets of special rubber attached to the underside of the covers by flat steel bars and studs. It is claimed that Rolltite covers achieve higher standards of reliability and total weathertightness than have hitherto been possible in steel hatch covers.

5.5.7 Drive systems

The drive mechanism is built into one of the stowage drum supports. Normally it is electrically powered but hydraulic powering can be fitted.

5.5.8 Weight

Rolltite covers are 20–30% heavier than single pull covers for the same hatchway as shown in Fig. 7.5.

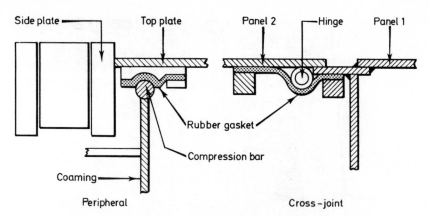

Fig. 5.36. Flexible rubber cross-joint seals for roll-stowing covers remain permanently attached at all stages of operation.

5.5.9 Application

Although roll-stowing covers can be used on the weather decks of most ships, installations to date have been in general cargo ships and bulk carriers. They are extremely simple to operate, quiet, very reliable and weathertight. As yet they are not very numerous. Their first cost is higher than that of the wire-operated standard single pull cover, although single pull covers fitted with additional features such as automatic lifting and cleating, which are inherent in the Rolltite design, would cost almost the same, as indicated in Table 7.1.

5.6 Side and end rolling covers

5.6.1 Description

Rolling covers usually consist of two large panels at each hatchway. They are fitted with wheels which roll along a track at both sides of the coaming top. Stowage rails, which may be portable, extend this track via pillars welded to the deck. In some installations, the wheels are not attached to the hatch cover but to the coaming and to fixed pillars on the deck, and the cover rolls across them. Apart from stowage location, the principal difference between side and end rolling covers is that the joint between side rolling panels is longitudinal and between end rolling panels is athwartships. Fig. 5.37 illustrates a typical arrangement which could be either side or end rolling; Fig. 2.7 illustrates the former.

These covers are usually fitted to large ships. They are often extremely heavy owing to their large dimensions and require hydraulic pot lifts (rams) to raise them into the rolling position. These hydraulic lifts are fitted to the

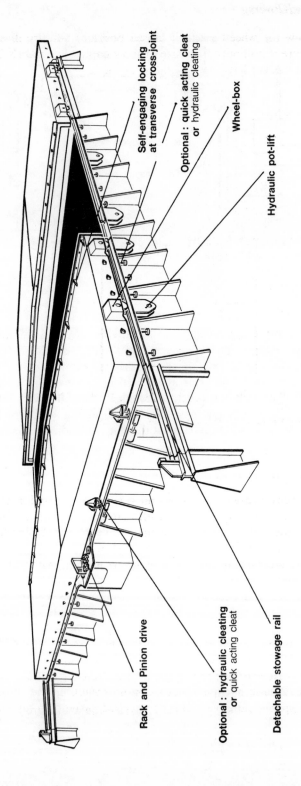

Self-engaging locking
at transverse cross-joint

Optional: quick acting cleat
or hydraulic cleating

Wheel-box

Hydraulic pot-lift

Rack and Pinion drive

Optional: hydraulic cleating
or quick acting cleat

Detachable stowage rail

Fig. 5.37. A pair of typical side-rolling covers with rack and pinion drive and hydraulic lifting and cleating.

coaming below the wheels (in their closed position) and are illustrated in Fig. 5.38. There is no limit to the size of these covers, and panels 20 metres square (20 m × 20 m) have been installed in ships.

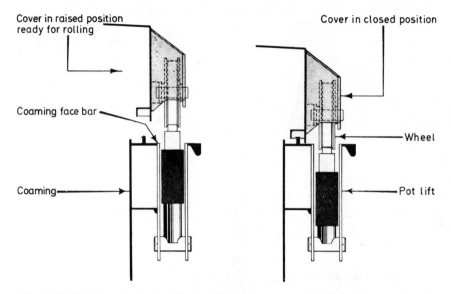

Fig. 5.38. Details of a hydraulic pot-lift. These are attached to the coaming immediately beneath the cover wheels when the hatch is closed. The rams raise or lower the wheels through slots in the coaming bar.

5.6.2 Operation

Before rolling covers are opened, they must be uncleated and lifted into the rolling position by the hydraulic pot lifts already mentioned. The rolling mechanism is then set in motion. This may move the panels simultaneously or singly clear of the hatchway and onto the stowage rails. When fully open the panels are stopped by the operator or by a limit switch fitted to the pillar at the end of the stowage rail. Closing the covers is carried out by reversing the foregoing process.

5.6.3 Construction

Construction is once again of the stiffened-plate type, usually with peaked tops which saves weight and improves water shedding. Unlike single pull covers the peak ridge of side rolling covers runs transversely. The beams and stiffeners are usually massive, because most rolling covers have a large unsupported span. The covers can be supplied plated top and bottom if required, to form a double skin, but in combination carriers there can be the danger of an explosive atmosphere forming in the void space.

5.6.4 Stowage

These covers neither tip nor fold, but roll horizontally to expose the hold. Thus they usually require at least one half of the total hatchway width at each side or end of the hatch opening for stowage. Hatches with side rolling covers are thus usually less than 50% of the breadth of the ship, unless covers overhanging the side of the ship when in the open position are acceptable. Where hatches are less than 30% of the breadth, as in ore carriers, single panel covers may be fitted. The coaming height on large bulk carriers is sometimes arranged so that it is possible to walk underneath the open covers when going from end-to-end or side-to-side of the ship.

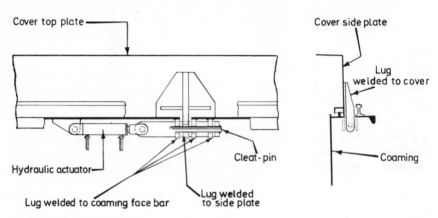

Fig. 5.39. Hydraulic peripheral cleating for side or end-rolling covers.

5.6.5 Cleating

Screw or quick-acting cleats may be employed around the hatch perimeter as described in Section 5.2.5 Automatic cleating is also available, usually by means of hydraulically actuated pins which are inserted into holes drilled in lugs, welded to the coaming or to the cover as illustrated in Fig. 5.39. The cross-joints are secured by pins fitted to the end of one panel which enter holes in the end of the other panel as the two come together on closing, as shown in Fig. 5.40.

5.6.6 Weathertightness

The usual systems of seals and compression bars, such as those illustrated in Fig. 5.21, are employed to make rolling hatches weathertight. But, since

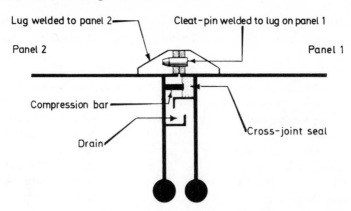

Fig. 5.40. Cross-joint cleating and sealing arrangements for side or end-rolling covers. Note that the face bars of the webs are of round section to allow ease of cleaning.

rolling covers are often fitted in large ships with large hatchways where distortion may be a problem, other sealing methods are sometimes used. These are designed to take up the relative displacement of cover and coaming and are particularly necessary in oiltight installations, as in combination carriers. An example is illustrated in Fig. 5.41 where a steel frame containing a gasket is compressed by a bar held down by springs. It is attached to each cover panel by a flexible rubber skirt with special 'grippers' fitted to the coaming to prevent the cover being lifted off.

5.6.6.1 *Variations arising from stowage considerations*
Single side rolling covers are sometimes fitted in ore carriers, rolling only to one side. This enables the total height of the ship above the waterline on one side (the air draft) to be reduced to allow the ship to pass under an item of shore-based cargo handling equipment if necessary. A further variation of this is the two-panel cover, where both covers can stow together to one side or the other, leaving only half the hatchway open, as well as in the more usual arrangement of one panel on each side.

5.6.7 Drive systems
Drive mechanisms for rolling covers have to propel the panels into stowage, even against an adverse heel (as much as 5° in combination carriers) or trim. They must also prevent panels rolling out of control down an incline resulting from heel or trim.

5.6.7.1 *Wire and chain drives*
In this arrangement wires or chains attached to the covers are rove through

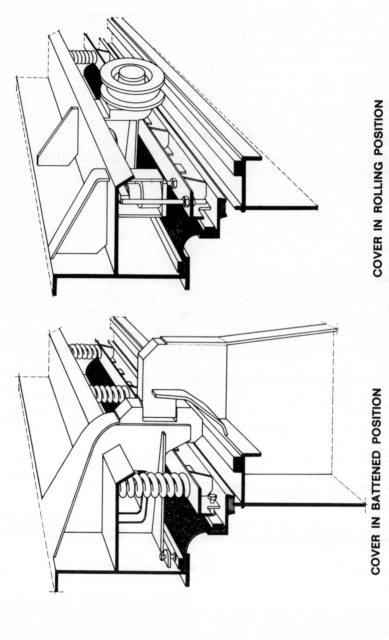

COVER IN BATTENED POSITION   **COVER IN ROLLING POSITION**

Fig. 5.41. Navire Hydroseal – an arrangement in which a non-rigid steel frame is attached to the main cover via a flexible rubber skirt to provide a weathertight seal which can absorb any flexing of the ship and distortion of the coaming.

blocks and led to an electric or hydraulic winch by which each panel is hauled into the open or closed position.

5.6.7.2 *Rack and pinion drive*

This is by far the most common means of moving the panels and is illustrated in Fig. 5.42. It provides very accurate position control and is usually powered by a low-speed hydraulic motor. The rack is divided at the perimeter seal to allow it to pass across the seal from the outside to the inside of the cover and the drive must therefore be transmitted by two pinions. This allows a wider gasket to be employed than is possible with a continuous rack where the pitch of the teeth limits its width. The rack is of the open bottom type and is therefore self-cleaning.

5.6.7.3 *Traction drive*

This is illustrated in Fig. 5.43 and is used for moving exceptionally heavy hatch cover panels. It consists of an electric motor, reduction gearbox, robust bearing blocks and a drive wheel. Two or four such units are usually necessary for each panel. They are enclosed in watertight compartments within the panel, and centrally controlled.

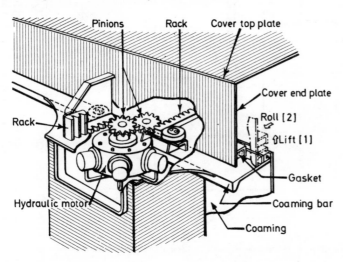

Fig. 5.42. Rack and pinion drive for rolling covers. The pinion is usually hydraulically powered.

5.6.8 Weight

The approximate weight of these covers is indicated in Fig. 5.44. The additional weight for stowage rails at the end of the coamings is from 2 to 3 tonnes per hatch end.

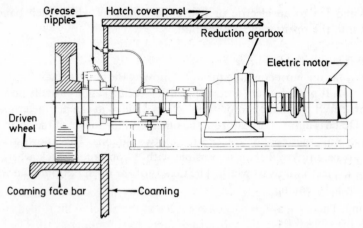

Fig. 5.43. Traction drive system for large, heavy rolling covers.

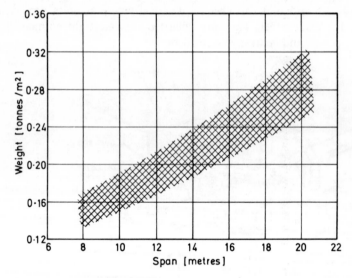

Fig. 5.44. Rolling cover weights per square metre as a function of span. Oiltight covers are at the heavier end of the range shown.

5.6.9 Applications

Rolling covers can be used on the weather decks of most types of ship, but are most commonly found in large bulk carriers. Since they can be made oil-tight, they are also very suitable for combination carriers. These covers are simple in both construction and operation. They are easy to maintain and are thus extremely reliable. Their principal drawback is that their stowage requirements limit the size of hatchway which can be accommodated in a given ship length and beam. The length of hatchways in ships fitted with end rolling covers is also limited because of the need for sufficient room to stow the panels, but in practice most rolling covers are side rolling.

5.7 Lift and roll covers (piggy-back)

5.7.1 Description

Lift and roll covers (often called 'piggy-back' covers) are a development of rolling covers and an example is illustrated in Fig. 5.45. Each cover consists of two panels, one of which has powered wheels. In the way of the 'dumb' panel, four hydraulic rams which act vertically upwards are fitted to the coaming. These engage in lugs on the sides of the panel to lift it high enough above the coaming for the motorized panel to roll underneath. The dumb panel is then lowered onto the motorized panel so that both can be moved together. The side coamings can be extended so that the two panels can be stowed beyond the hatch end. Alternatively these covers can be side-rolling,

Fig. 5.45. Lift and roll (piggy-back) covers on a forest products carrier. A 'dumb' panel is shown resting on top of a powered panel. The two gantry cranes span the full width of the hatches.

stowing abreast the hatchway. There is in theory no limit to the size of these covers; covers composed of two 100-tonne panels for 26 m by 23 m hatchways have been supplied for forest product carriers.

5.7.2 Operation

The covers are first uncleated, manually or automatically, and then lifted hydraulically. The dumb panel is raised by the four rams already mentioned and the motorized (wheeled) panel is lifted for rolling by means of pot lifts as illustrated earlier in Fig. 5.38. The wheeled panel rolls under the dumb panel which is then lowered onto it. The two panels may be left in this condition, with the hatch opening partially exposed, or they may be rolled clear of the hatchway.

5.7.3 Construction

Like other hatch panels, a stiffened-plate construction is adopted for lift and roll covers. They are invariably flat-topped, both to facilitate the loading of deck cargo and so that the dumb panel may sit firmly on its motorized supporter. Double-skinned covers are available if desired.

5.7.4 Stowage

If it is not necessary to open the whole hatch, these covers can be stowed at one end of the coaming, and thus no deck stowage space is required. If however, they must be removed clear of the hatchway, stowage space having one half of the hatchway area and twice the panel depth is necessary. Coaming heights are sometimes arrranged to allow personnel to walk under covers when they are stowed on rails, as shown in Fig. 5.46.

5.7.5 Cleating

Peripheral cleating can be by screw or quick-acting cleats as described in Section 5.2.5. Automatic cleating actuated by hydraulic cylinders at each side of the hatch is also available. This is similar to the cleat illustrated in Fig. 5.11. The cylinders are connected to sliding bars which cause the cleat hooks to rise and rotate to engage lugs welded to the cover side plates. Cross-joint cleats are built into the panels and engage as the panels come together.

5.7.6 Weathertightness

Lift and roll covers are made weathertight in the same way as rolling covers, except that the flexible rubber and steel skirt described in Section 5.6.6 is not used.

5.7.7 Drive systems

The motorized panel can be driven in a variety of ways, as described below.

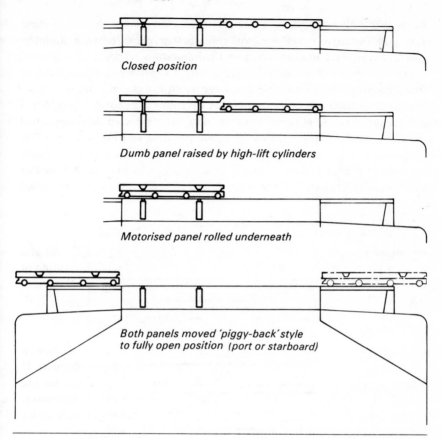

Closed position

Dumb panel raised by high-lift cylinders

Motorised panel rolled underneath

Both panels moved 'piggy-back' style
to fully open position (port or starboard)

Fig. 5.46. Operation of a lift and roll cover installed transversely. Note that only the wheeled panel is powered.

5.7.7.1 Fixed and long chain drives
The chain drive mechanisms are similar to those described in Sections 5.2.7.1 and 2 for single pull covers.

5.7.7.2 Rack and pinion drive
This drive is identical to that fitted to rolling covers and is described in Section 5.6.7.2.

5.7.8 Weight

The weight of these covers is similar to that of side and end rolling covers described in Section 5.6.8.

5.7.9 Applications

Lift and roll covers are used on the weather decks of bulk and forest product carriers, container ships and combination carriers. They have also been used for flush tween-deck covers. Fig. 5.47 shows an installation on a multi-purpose vessel which has one large hatchway closed by four panels; all the possible combinations of hatchway openings are shown. With fore and aft rolling covers, adjacent hatches can be connected by rails so that individual panels can be rolled from one hatch to another.

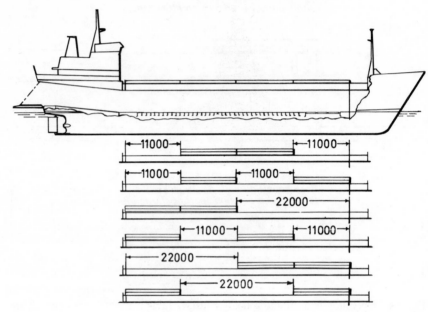

Fig. 5.47. Various opening arrangements for lift and roll covers fitted to a single long hatchway. Sliding covers are fitted at the tween-deck.

5.8 Sliding tween-deck covers

5.8.1 Description

Sliding tween-deck covers are made up of a number of Z-section plates which slide under each other to expose the hatchway. A typical installation is illustrated in Fig. 5.48 where half a cover is shown in the process of being opened while the other half is in the closed position. When closed the covers are flush with the surrounding deck, thereby facilitating the use of fork lift trucks. They can support normal cargo loads and can be fitted with container locating points if sufficiently strengthened. These covers are normally not used on tween-deck hatchways wider than 16 m.

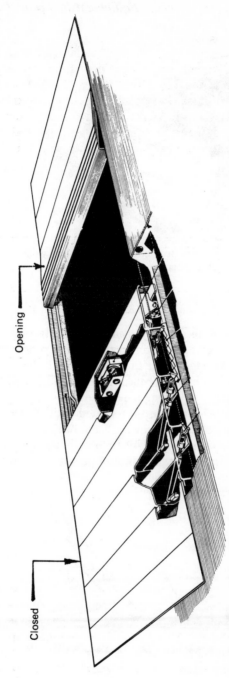

Opening

Closed

Fig. 5.48. Sliding tween-deck covers showing the drive unit.

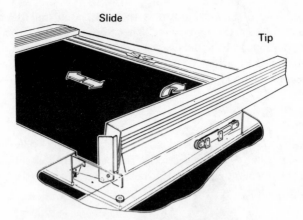

Fig. 5.49. Sliding tween-deck covers can be tilted when open to provide the greatest possible clear opening.

5.8.2 Operation

These covers are not secured by cleats, and their operation is entirely automatic. During opening a wheel on the leading panel moves up a ramp, which tips its trailing edge under the second panel; the leading panel then moves back until it lies entirely beneath the second panel. When this stage has been reached, the two leading panels move together until they are both under the third panel. This process is repeated until all the panels are nested together at the end of the hatchway, from which position they can be folded back as shown in Fig. 5.49 to obtain the largest possible clear opening. To close the cover, the nested panels are first tipped into the

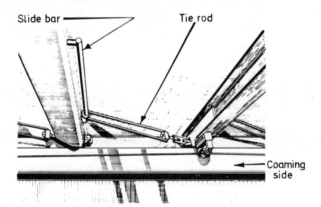

Fig. 5.50. Underside of sliding tween deck-covers showing the tie rods and slide bars.

horizontal position. The drive mechanism is then activated and the nest moves towards the centre of the hatchway. When the panel closest to the stowage end is fully deployed, the panel next to it emerges from the nest. This operation is accomplished by means of two sliding rods, one at each side of the panel as shown in Fig. 5.50 and is repeated until the cover is completely closed.

5.8.3 Construction

Individual panels are constructed of heavier-gauge flanged steel plate than most of the covers described in earlier sections. They are shaped as shown in Fig. 5.51. Because the panels stow underneath each other, their vertical web provides the only stiffening.

5.8.4 Stowage

Table 5.3 summarizes the stowage requirements for sliding tween-deck covers for various hatch lengths and gives the number of panels required.

Table 5.3. Stowage lengths L and numbers of panels N for various sizes of hatch opening S
(all dimensions in mm)

L	N	S	L	N	S	L	N	S
2 330	2	1 100	8 210	10	1 900	14 090	18	2 700
3 065	3	1 200	8 945	11	2 000	14 825	19	2 800
3 800	4	1 300	9 680	12	2 100	15 560	20	2 900
4 535	5	1 400	10 415	13	2 200	16 295	21	3 000
5 270	6	1 500	11 150	14	2 300	17 030	22	3 100
6 005	7	1 600	11 885	15	2 400	17 765	23	3 200
6 740	8	1 700	12 620	16	2 500	18 500	24	3 300
7 475	9	1 800	13 355	17	2 600			

*This value varies between 400 mm and 1400 mm and can be altered to suit the hatch length

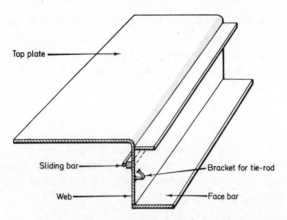

Fig. 5.51. Construction of a sliding tween-deck cover panel.

5.8.5 Cleating and weathertightness

As these covers are only used in tween-decks, they need not be weathertight and do not require cleats.

5.8.6 Drive system

An electric winch unit is built into a compartment within the leading panel as illustrated in Fig. 5.52. Fixed wires are attached to the ends of the hatchway and run along each side coaming, and round each winch drum. As the drum turns, so they and the leading panel move along the wire. Rack and pinion drive is, however, now increasingly preferred.

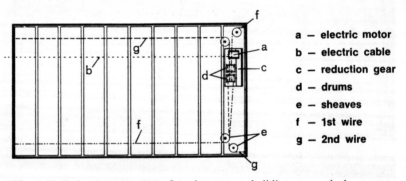

Fig. 5.52. Drive arrangement of a wire-operated sliding tween-deck cover.

5.8.7 Weight

Weight is comparable with other types, typically 0·20–0·25 tonnes/m², depending on span and loading.

5.8.8 Applications

These covers can only be used in tween-decks and are suitable for all non-refrigerated general cargo and multi-deck vessels. They can also be used to provide an extra deck within the weather deck coaming as illustrated in

Fig. 5.53. Sliding tween-deck cover in service here to allow separate use of the weather deck coaming space, e.g. for cars as illustrated in Fig. 6.59.

Figs. 5.53 and 6.59 – a useful means of gaining extra deck space if required, for example, for carrying cars.

5.9 Pontoon covers

5.9.1 Description

Pontoons are the simplest form of steel hatch cover and are merely lifted on and off the hatch coaming. Two types are generally fitted;
  (i) single piece covers for weather decks particularly on cellular container ships;
  (ii) multi-panel covers on tween-decks of simple multi-deck cargo vessels.
Single piece covers are large and awkward to handle but since they can be lifted off the ship completely, no on-board stowage space need be provided. Often, they can be stowed on top of adjacent hatch covers. This is an advantage in large container ships where the non-opening deck space must be kept to a minimum. Hatchways can be covered by a single panel or by a series of interlocking panels. They are used for both weather and tween-deck

Fig. 5.54. Pontoon weather deck covers fitted in a LASH vessel. Pontoon covers are also shown in Figs. 8.7 and 10.5. Note the 500-tonne gantry crane.

hatchways. Tween-deck pontoons are usually in several sections so that they may be lifted through the weather deck hatchway and stowed on deck. The size of pontoons is limited by the capacity of the lifting gear available to remove them. In container ships the weight of each panel is usually less than 30 tons – the capacity of typical container cranes. Pontoon covers which are made weathertight by means of tarpaulins and battening devices, as described in Sections 4.1.5 and 9.1.1 are now rarely fitted.

5.9.2 Operation

Once uncleated, pontoon covers are removed by lifting tackle and landed on deck, on the quay or, in the case of buoyant pontoons, over the side into the water. Pontoons can be lifted by shore cranes or ship's gear. In container ships they are fitted with container corner fittings so that they can be handled with a standard container spreader, as well as with suitable corner fittings and lashing points for deck containers, as shown in Figs. 8.7 and 10.5.

5.9.3 Construction

Stiffened plate construction is employed for pontoon covers; buoyant pontoons are constructed as watertight boxes. Higher tensile steel is often used in pontoon covers on container ships to reduce their weight. Pontoon covers, especially those used in tween-decks, are sometimes constructed with sides angled inwards so that they fit like plugs into the coamings.

5.9.4 Stowage

As already mentioned above, no stowage space is required if the pontoons are lifted off the ship or can be stowed on top of adjacent covers. Where space permits, pontoon covers can be stowed on deck abreast the hatchways.

5.9.5 Cleating

Standard screw or quick-acting cleats described in Section 5.2.5 are normally employed with pontoon covers. Tween-deck pontoons are fitted with the 'token' screw-down cleats described in Section 5.3.5 or with a combined type of self cleating or hook/lifting eye.

5.9.6 Weathertightness

Gaskets, compression bars and drainage channels similar to those described in Section 5.2.6 are normally used to secure pontoon covers, which means that an opening sequence usually has to be followed. Tween-deck pontoon covers are not usually weathertight, and can therefore be opened in any order.

5.9.7 Drive systems

Pontoon covers have no drive system.

5.9.8 Weight

Single panel pontoon covers are usually about 10% lighter than the equivalent single pull covers shown in Fig. 5.16, with tween-deck pontoon covers weighing about 0·24–0·28 tonnes/m², when designed for container or wheel loads.

5.9.9 Applications

Pontoon covers can be used for both weather decks and tween-decks. They are generally found on the weather decks of container ships, barge carriers and their lighters and the tween-decks of general cargo vessels. They are also increasingly used in specialized heavy-lift ships as they can be made extremely strong. Great Lakes ships also often have them and are sometimes equipped with special pontoon cover lifting cranes. They may be made completely flush with surrounding decks. They are also easily insulated and hence can be used on the tween-decks of refrigerated ships.

5.10 Telescopic covers

5.10.1 Description

Telescopic covers designed for barges and inland waterway craft usually consist of sets of two or three panels shaped as shown in Fig. 5.55 with each panel slightly smaller than the one before it. Each panel has wheels on both sides with those on one side guided by a rail while those on the other side are unrestrained to allow for any displacement or twisting of the hatch coaming.

5.10.2 Operation

To open telescopic covers, they must be uncleated, nested together, and rolled by hand to any position along the length of the hatchway. They are secured in the open position by means of locking devices fitted to the wheels of each set of panels. By this means up to 65% of the total hatchway length can be open at the one time.

5.10.3 Construction

The covers consist of a steel frame of longitudinal and transverse girders to which are attached wheels and cleating devices. A lightweight skin of aluminium alloy is rivetted to the frame making it weathertight and light enough to be rolled by hand. The panels can support the weight of a man walking on them, but not deck cargo.

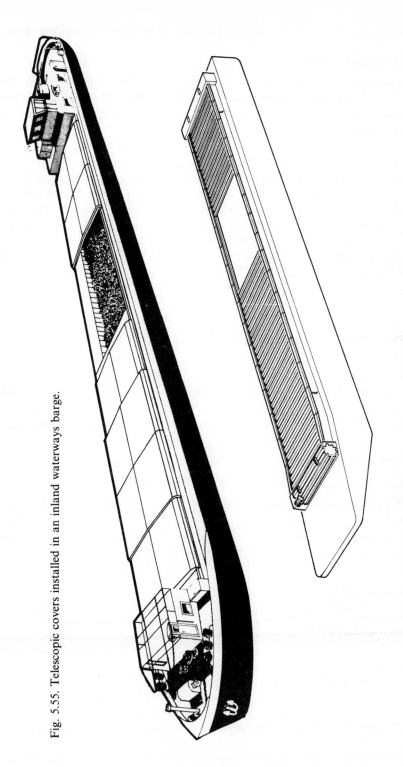

Fig. 5.55. Telescopic covers installed in an inland waterways barge.

Fig. 5.56. Flexible rolling covers installed in an inland waterways barge.

5.10.4 Stowage space

No additional stowage space is required as the covers always remain over part of the hatchway, taking up at least 35% of the hatchway length.

5.10.5 Cleating

The covers are secured to the coaming by means of self-tensioning locks at the ends of each set of panels. Cross-joints between the panels are secured by flanges on the end girders which overlap when the covers are closed.

5.10.6 Weathertightness

The hatch covers on inland waterway craft are only intended to be rain-tight and hence do not require gaskets and sealing bars. The covers are designed to overlap the coaming, thereby ensuring that rain does not enter the hold.

5.10.7 Applications

Telescopic covers are restricted to barges and inland waterway craft and have the advantage of being easy and quick to operate. They can be fitted in sets of two panels where the maximum clear opening at any one time is 50% of the hatchway length, or in sets of three panels where the maximum opening is 65% of the hatchway length.

5.11 Flexible rolling covers

5.11.1 Description

Flexible rolling covers consist of a one-piece, transversely corrugated, flexible alloy sheet which rolls onto a stowage drum at one end of the hatchway. Two units, each stowing at opposite ends of the hatchway and meeting in the middle can be fitted to each hatch.

5.11.2 Operation

To open a flexible rolling cover, the side panels are swung down and the cover is wound onto its stowage drum either by hand using a crank, or by an electric motor. The reverse operation closes the covers.

5.11.3 Construction

The construction of flexible covers is illustrated in Fig. 5.56. They are extremely light as they are primarily a means of protection against rain and pilferage.

5.11.4 Stowage space

The diameter of the stowage drum depends on the length of the hatchway, but for a typical 1300 tdw European barge it is about 1·0 m with a drum at each end of the hatchway.

5.11.5 Cleating and weathertightness

Glass reinforced plastic (GRP) side panels, such as are shown in Fig. 5.57, have sponge rubber inserts to provide protection against rain. Like telescopic covers, flexible covers are only required to prevent the entry of rain water. The side panels can be in any of three positions, namely,

(a) closed, where they provide a barrier against the weather,
(b) tipped when the covers may be opened,
(c) swung away completely to give unimpeded access to the hold.

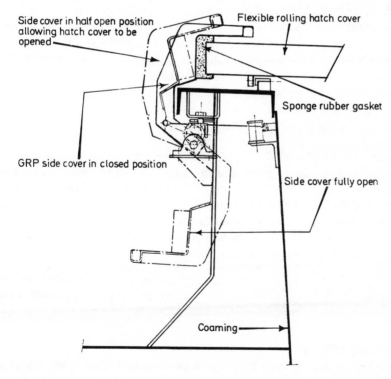

Fig. 5.57. Section through the coaming of a barge hatchway arranged for a flexible rolling cover. Note the spray-tight GRP side panels.

5.11.6 Drive systems

Flexible covers can be moved manually be means of a crank or they can be powered by an electric motor and gearbox.

5.11.7 Applications

These covers are restricted to barges and inland waterways craft. They are simple and quick to operate, require little stowage space, and provide a completely clear hatch opening. However, they are very light in construction and are not designed to support the weight of any deck cargo or even a man.

In practice, very few barges have been fitted with modern types of hatch cover. Their operators have generally preferred to continue to use the cheaper beams and boards because rapid opening and closing is less important in inland waterway vessels, which often lie alongside a quay for long periods, than in ocean-going vessels where turnround times are critical.

5.12 Miscellaneous covers

5.12.1 Deep tank covers

Many multi-deck general cargo ships have spaces that can be used either for cargo or for additional water ballast. Often they are lower holds near midships, made into a watertight space by means of watertight hatch covers (see Fig. 2.9). This extra ballast space is known as a deep tank, and may be up to 2000 tonnes capacity. In some ships deep tanks are used to carry liquid cargoes. The simplest type of deep tank cover consists of a stiffened steel plate sealed by a rubber gasket and secured with studs and nuts. Others have eccentric wheels and cleating systems like those described in Section 5.2.2. The covers are usually fitted flush or to a low coaming about 150 mm high.

5.12.2 Beams and boards covers

The traditional covers of steel beams and wooden boards are rarely fitted today, except occasionally on tween-decks where tarpaulins are unnecessary and cheapness and ease of partial opening by hand for occasional use may outweigh their disadvantages. Fig. 8.21 shows a heavy-lift ship with beams and boards covers fitted on the tween-deck. The general details of such covers are as described in Chapter 1, except that welded construction is now used.

5.12.3 Tank hatches

Small hatches are fitted on the weather decks of tankers to allow personnel to enter tanks for inspection and tank cleaning purposes. Broadly similar personnel access hatches are installed over the upper wing tanks of some bulk carriers where these spaces are used for the carriage of grain. A typical hatch is illustrated in Fig. 5.58.

Fig. 5.58. A simple tank hatch, for personnel access and tank cleaning, 760 mm wide.

CHAPTER SIX

Access Equipment for Horizontal Loading Ships

Summary

This chapter examines the roll-on/roll-off concept and goes on to describe its implications for ship design in general and access equipment design in particular. Ro-Ro cargo handling equipment is described with special emphasis on container handling, and the principal characteristics of vehicles found in Ro-Ro vessels are listed. The chapter concludes by describing the variety of Ro-Ro access equipment available to the ship designer: ramps, bow and stern openings, elevators, side doors, bulkhead doors and car decks.

6.1 Design philosophy

Horizontal-loading vessels are commonly referred to as roll-on/roll-off or Ro-Ro ships and accept their cargo on wheeled vehicles via openings in the bow, side or stern. Not all horizontal-loading ships are strictly Ro-Ro ships however. For instance, pallet carriers may have side doors only, and accept cargo from fork lift trucks operating on the quayside (see Section 2.6.6).

Train ferries apart (see Section 8.3.4.2), the first purpose-built deep-sea Ro-Ro ship was the USNS *Comet* which was built in 1958, for the United States Military Sea Transportation Service, to carry military vehicles across the Atlantic with as quick a turnround as possible.

Fig. 6.1 shows a hypothetical ship fitted with all the usual forms of Ro-Ro equipment, showing what is available and how it is used.

6.1.1 Traffic movements

Flexibility and fast port turnrounds are the essential features of Ro-Ro operations. High cargo handling speeds are therefore vital. Cargo handling rate can be increased in three ways:

(i) by increasing the unit load per vehicle until the vehicle's size impairs its manoeuvrability and speed;

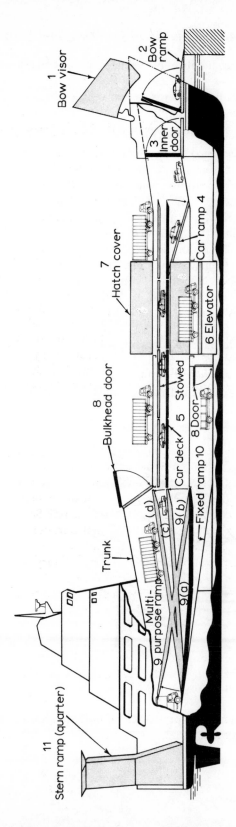

Fig. 6.1. The range of access equipment available for **Ro-Ro** ships is illustrated by this hypothetical vessel. 1. Bow visor (Section 6.4.1) 2. Bow ramp (Section 6.3.9.4) 3. Inner bow door (Section 6.3.10.2) 5. Car ramp (Section 6.3.9.4) 4. Car ramp (Section 6.8) 6. Elevator (Section 6.5.2.) 7. Hatch cover 8. Bulkhead door (Section 6.7.2) 9. Multi-purpose ramp showing alternative positions (Section 6.3.10.4) 10. Fixed ramp (Section 6.3.10.1) 11. Stern ramp and door (Section 6.3.9)

Bow visor 1

2 Bow ramp

3 Inner door

Car ramp 4

7 Hatch cover

6 Elevator

5 Stowed

Car deck 5

8 Door

Fixed ramp 10

8 Bulkhead door

Trunk

Multi- 9 purpose ramp

9(d)

9(b)

9(a)

11 Stern ramp (quarter)

(ii) by increasing the number of vehicles until congestion impairs throughput;

(iii) by changing traffic flow patterns so that vehicles do not have to make tight manoeuvres (for example, by providing both stern and bow access so that 'U' turns are unnessary).

Roll trailers, sometimes called Mafis after a major manufacturer, are often used for transferring containers into and out of Ro-Ro ships. They are low vehicles, unsuitable for road use, which are towed by terminal tractors, sometimes called Tugmasters after a particular manufacturer's design, and can carry containers singly, or, if greater throughput is necessary, in pairs stacked one above the other. The latter arrangement is only possible in ships having at least 6 m clear headroom in the vehicle deck space – sufficient to accommodate two 8 ft 6 in (2·59 m) containers. Another arrangement, capable of transporting four 20 ft or 40 ft containers has also been developed in recent years in response to the need for higher handling rates although it is not widely used. This is known as the Lifting Unit Frame (LUF) system and is illustrated in Fig. 6.4(d).

Conventional Ro-Ro terminal equipment, such as tractors, trailers, fork lift trucks and straddle carriers, can be increased in number to raise the cargo handling rate until queueing problems restrict throughput. Often the rate of vehicle arrivals exceeds the rate at which cargo can be stowed or

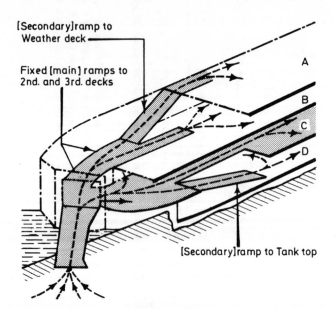

Fig. 6.2. Traffic flows in a Ro-Ro ship. The density increases towards the stern ramp.

lashed. It is therefore common for more vehicles to be employed discharging a vessel than loading it, because unlashing and breaking down the stow is generally faster than stowing and lashing. Typically, a small Ro-Ro vessel, working one deck at a time, might employ five tractors for discharging and three for loading.

Consider a Ro-Ro ship with one stern ramp and two main internal vehicle ramps, each fed by secondary ramps, as illustrated in Fig. 6.2. If decks A, B, C and D, or any combination of them, are loaded or discharged simultaneously, it is clear that the traffic density increases towards the stern ramp. Thus the cargo handling rate depends on the rate at which vehicles can cross the stern ramp. This means that the ramp arrangement must be carefully designed to ensure that the system is capable of its intended throughput.

The third way to increase cargo handling rate, e.g. designing the vessel for one-way vehicle traffic, is only suitable for specialized trades and requires special port facilities if extensive access equipment is not fitted. Vessels operated in this way include passenger/vehicle ferries equipped with axial bow and stern ramps. They often occupy corner berths as shown in Fig. 6.3. This arrangement is most suitable for the carriage of trailers and other vehicles that are parked with their power units attached so that they can be driven straight off on arrival at their destination. Terminal tractors cannot reach parked trailers from both ends of the parking lane if two-way traffic is prohibited.

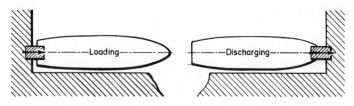

Fig. 6.3. Berthing arrangements for a passenger/vehicle ferry equipped for drive-through traffic.

6.1.2 Cargo handling methods

Ro-Ro ships are used to carry wheeled or tracked vehicles, bulky items on trailers, containers and flats, pallets, and block-stowed cargo which does not travel on its own wheels, for example forest products. A typical cargo might consist of a combination of all of these.

Wheeled and tracked vehicles do not usually present design or operational problems unless axle loads are very high, but provisions (such as planks of wood) must be made to prevent tracks damaging the deck. In some cases operators have laid temporary rails on stern ramps and decks for locomotives and rolling stock.

Table 6.1. Characteristics of typical vehicles used on Ro-Ro vessels

Equipment	Length, metres	Width, metres	Unladen height, metres	Height loaded with ( ) high containers metres	Tare weight, tonnes	Max. payload, tonnes	Gross weight, tonnes
25-tonne fork lift truck	9·0	3·2	3·2	4·4 (1)	32	25	57
4-tonne fork lift truck	4·0	1·3	2·2	—	6	4	10
Side loader	9·0	3·7	4·3	4·6 (1)	35	25	60
C-van	14·0	4·7	2·8	3·3 (1)	22	30	52
Straddle carrier	6·0	3·9	4·5	4·5 (1)	18	22	40
Rigid lorry	11·0	2·5	4·0	—	8	16	18–24
Articulated lorry	15·0	2·5	4·0	—	11	21–27	32–38
Motor car	4·0	1·6	1·4	—	1·5	0·5	2
Terminal tractor	4·8	2·5	2·6	—	7·2	80 (tow)	—
LUF for 8 TEUs with Lufmaster	18·8	4·9	2·6	5·9 (2)	67	180	247
LUF for 4 TEUs with terminal tractor	10·5	6·1	2·6	5·8 (2)	25	100	125
20 ft roll trailer	6·1	2·5	0·6	3·4 (1)	2	20	22
40 ft roll trailer	12·3	2·5	0·8	3·6 (1)	5	40	45
40 ft road trailer, flat bed	12·2	2·5	1·4	3·9 (1)	5	27	32
40 ft road trailer	12·2	2·5	4·0	—	6	24	30

*Can be part of ship's own equipment

Bulky machinery and structural items also cause few problems. They are usually placed on road or roll trailers and towed aboard the ship by tractors, which leave them in their stowed position ready for lashing. At their destination they are discharged by another terminal vehicle.

Containers can be handled in a variety of ways. The method chosen depends on the number and size of containers to be handled. On well-established routes between terminals equipped with container cranes and having a large volume of container traffic, hybrid vessels, with cellular as well as Ro-Ro capacity, are often used. This is discussed in Section 8.3.3.2. In other Ro-Ro types containers are usually tightly block-stowed, both below decks and on the weather deck. A tight block-stow is one in which the containers are stacked fore and aft or athwartships and lashed one or two high, and as close together as possible.

Max. axle load (laden) tonnes	Number of axles when moving	Max. speed laden on level road (km/h)	Stowed on board with cargo	Remarks
54	2	30	no*	With 20 ft container 20 tonne capacity Front axle 4 wheels Length includes 2·5 m forks
8	2	30	no*	Length includes 1·2 m forks
30	2	30	no*	Model for 20 ft or 30 ft container
13 × 2	4	15	no*	Model for 40 ft container. 20 ft version available. 8 wheels on 4 independent axles
10 × 2	4	30	no*	Model for 20 ft container. 40 ft version available. 4 wheels, each on independent axle
10	2–4	80	yes	
10	3–5	80	yes	
1	2	100+	yes	Typical European saloon
25	2	35	no*	Gooseneck excluded
48	5	15	yes	8 wheels per 4 axles on bogie Length includes tractor with 1 axle
48	4	20	yes	8 wheels per 2 axles on bogie Length includes tractor with 2 axles
14	1	20	yes	With 20 ft container but excludes tractor. 5th wheel load 9 tonnes
32 (2 axles)	2	20	yes	With 40 ft container but excludes tractor, about 5·5 m extra length. 5th wheel load 15 tonnes Axles spaced only 0·7 m
10	2	80	yes	Excludes tractor, about 2·8 m extra length
10	2	80	yes	Excludes tractor

Table 6.1 lists types of equipment commonly used to handle cargo on Ro-Ro vessels, together with their main characteristics.

6.1.3 Cargo handling methods in different types of ship

6.1.3.1 *Short sea Ro-Ro vessels*

Two entirely different methods of handling containers in these ships have grown up. In the first, all containers are carried on trailers. This method is common on very short routes where port time can be as much as 50% of the round-trip time, and delays must be minimized. Both road and roll trailers are used. Containers are transferred to roll trailers at the port terminal and tractors are used to stow them in the ship. This involves a two-way traffic flow with tractor/trailers travelling in one direction and tractors in the other. Articulated tractor/trailer units are sometimes transported with their drivers as passengers but these account for only a small proportion of the traffic on most routes.

In the second method, containers are brought on road trailers alongside or to the weather deck, where they are lifted by gantry crane and stowed either on deck or in container cells below deck in the usual manner.

6.1.3.2 *Passenger/vehicle Ro-Ro vessels*

These ships usually operate on regular routes carrying self-propelled vehicles with their owners/drivers travelling as passengers. Loading and discharging is extremely fast and lends itself to the one-way traffic flow with vehicles entering via a stern ramp and leaving via a bow ramp. Long-established terminals with bridge ramps (linkspans) commonly serve these vessels.

6.1.3.3 *Car carriers*

The cargo decks of pure car carriers have little headroom, usually about 1·65 m. Decks and ramps are designed for car weights of only 1 or 2 tonnes. These ships have side ramps and occasionally a stern ramp. One-way traffic flow is normally adopted since return cargo is uncommon in the export car trade. Unlike passenger/vehicle ferries, car carriers employ a team of stevedore drivers to load and discharge their cars.

6.1.3.4 *Deep sea Ro-Ro/container vessels*

These have cellular container capacity which is loaded and discharged by shore cranes in the normal way. Their Ro-Ro capacity is often used for non-containerized cargo such as bulky machinery, large vehicles and miscellaneous items such as boats, locomotives, helicopters etc. These are taken aboard on road or roll trailers towed by tractors. Ro-Ro container ships usually have car carrying capacity of low headroom, sometimes including narrow wing spaces which are unsuitable for most other cargo.

6.1.3.5 *Pure deep sea Ro-Ro vessels*

These ships are designed around unit load modules, usually containers, with additional containers stacked on the decks. The cargo handling equipment they commonly employ is listed in Table 6.2. Front loading fork lift trucks with containers carried transversely as shown in Fig. 6.4(f) make the tightest stow and are most effective when containers are fitted with 'pockets' for the fork lift tines. Containers without pockets can be moved on flats, Fig. 6.4(k), or trailers, Figs. 6.4(a) and (b). Fork lift trucks fitted with overhead or side spreaders (Figs. 6.5 and 6.6) are also suitable for containers without pockets. Most overhead types require about 1·2 m of clear deck height above the container stack, the side spreader about 0·3 m. Trucks often need to work in reverse for maximum visibility.

Table 6.2. Container handling equipment for Ro-Ro vessels

Equipment	Maximum container capability	Tight block-stow in ship	Maximum ramp slope	Typical deck height requirement for containers†	
				One high	Two high
Fork lift truck with tines (forks)	20 ft* (fitted with pockets)	yes	1 in 7	3·3 m	5·3 m
Fork lift truck with overhead spreader	20 ft*	yes	1 in 7	4·4 m	6·4 m
Fork lift truck with side spreader	20 ft*	yes	1 in 7	3·4 m	5·4 m
Roll trailer	40 ft	no	1 in 9	3·6 m (40 ft) 3·4 m (20 ft)	6·2 m (40 ft) 6·0 m (20 ft)
Road trailer	40 ft	no	1 in 6	4·0 m	—
Lifting unit frame	40 ft	no	1 in 9	3·2 m	5·8 m
Side loader	40 ft	yes (two high)	1 in 6	4·7 m	5·7 m
Straddle carrier	40 ft	no	1 in 9	4·6 m	—
C-van	40 ft	no (two high)	1 in 6	3·4 m	6·0 m

* Empty or lightly-loaded 40 ft containers can be carried but this is only done at container terminals
† Exact figures will depend on model and attachments selected and whether containers are 8 ft high, 8 ft 6 in, or higher

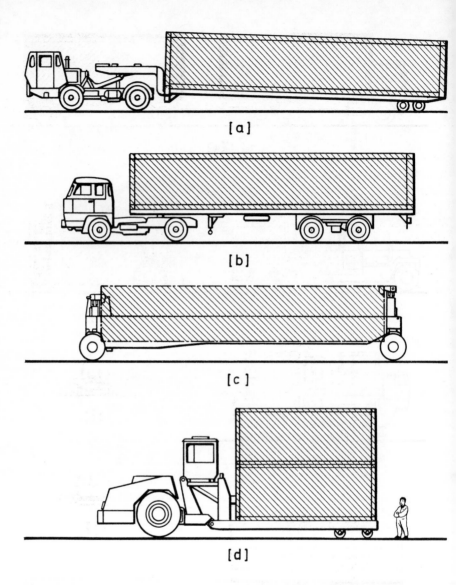

[a]

[b]

[c]

[d]

Fig. 6.4 (*See also opposite*) Vehicles and equipment found on Ro-Ro vessels, drawn to the same scale: (*a*) Terminal roll trailer for 40 ft containers, drawn by a terminal tractor fitted with a goose-neck drawbar. 20 ft trailer basically similar, but only one axle; (*b*) Articulated lorry for 40 ft containers; (*c*) C-van, so named because its plan view is in the shape of a 'C', used for transporting 40 ft containers; (*d*) Lifting Unit Frame (LUF) for four 20 ft containers with Lufmaster tractor. Trailers for four 40 ft containers are also available; (*e*) Rigid lorry with trailer; (*f*) Large fork lift truck for loaded 20 ft containers fitted with pockets for the fork lift tines; (*g*) Straddle carrier for 20 ft containers; (*h*) Side loader for 20 ft containers; (*i*) Small fork lift truck for light loads, pallets or loading containers; (*j*) Typical European saloon car; (*k*) Container flat fitted with tine pockets, which can be used for loaded 20 ft containers without pockets, or for miscellaneous cargo lashed to it.

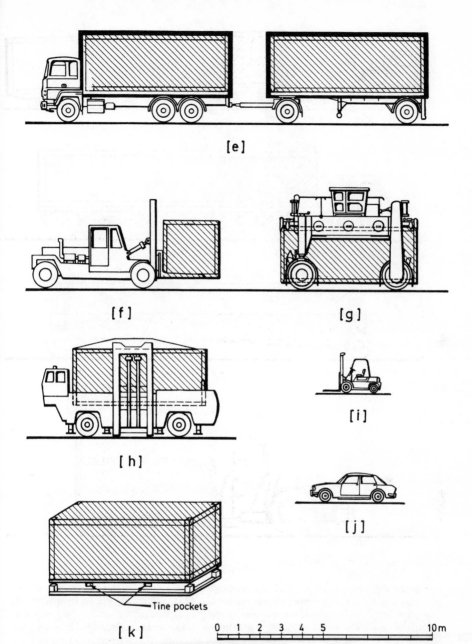

[e]

[f]

[g]

[h]

[i]

[j]

Tine pockets

[k]

0 1 2 3 4 5 10 m

Fig. 6.5. Fork lift trucks can be equipped to handle containers not fitted with fork (tine) pockets. The photograph shows a 25-tonne truck equipped to handle laden 20 ft containers with an overhead spreader. Note that the standard forks can be fitted into the spreader, this arrangement being suited to terminal operation, while a similar spreader is available which replaces the forks, thereby reducing the clear space required above the container for shipboard operation.

Fig. 6.6. Fork lift truck equipped with a side spreader to handle containers without pockets in minimum headroom. The photograph shows a loaded 20 ft container being carried aboard a large Ro-Ro via a quarter ramp.

The Lifting Unit Frame (LUF) can produce a fairly tight stow but some space is needed to accommodate the frame. It can be handled by a terminal tractor, or 'Lufmaster' (illustrated in Fig. 6.4(d)). LUF's advantage is that four or more containers can be moved in a single operation.

Side loaders, or side lifts, (Fig. 6.4(h)) can also stow tightly, except in the last lane that they work. They require more manœuvring space than a fork lift truck but offer better visibility when laden. Shipborne straddle carriers (Fig. 6.4(g)) and C-vans (Fig. 6.4(c)) do not make a tight stow owing to the space required for their wheels. The former can generally only stow one-high on ships, but both are able to carry loaded 40 ft containers, which is usually not possible with fork lift trucks. Most of the vehicles are powered by internal combustion engines, so special attention must be paid to ventilation and noise levels.

6.1.3.6 *Side loading ships*

These are loaded horizontally through side doors but do not normally accept cargo on wheeled vehicles. Palletized cargo is placed just inside the side door by a fork lift truck operating on the quay. It is then removed to its stowage position by shipboard fork lift trucks, elevators or conveyors. Fig. 2.5 shows a fork lift truck placing a pallet of cargo inside such a ship.

6.2 Ro-Ro access equipment

Fig. 6.1 shows the different types of specialized access equipment that Ro-Ro vessels use for horizontal cargo transfer. The various types of equipment are discussed in detail in the following sections.

6.3 Ramps.
6.4 Bow openings.
6.5 Elevators.
6.6 Side doors.
6.7 Bulkhead doors.
6.8 Car decks.

6.3 Ramps

6.3.1 General

Ramps are used externally to allow wheeled vehicles to travel between quay and ship, and internally to provide access from deck to deck and additional stowage area for cargo. They may be divided into two broad categories;

(a) external, bridging ship and quay;
(b) internal, linking adjacent decks.

External ramps can take several forms;

(i) axial (centre-line) bow and stern ramps;
(ii) angled quarter ramps, usually offset at an angle between 30° and 45° to the centre-line:
(iii) slewing ramps, usually mounted on the centre-line, that can be set to an appropriate angle up to about 40° from the centre-line on either side.

Internal ramps can take various forms;

(i) fixed;
(ii) movable, serving two or more decks, closing deck openings and providing additional stowage area.

6.3.2 Slopes or gradients

Ramp slope or gradient can be expressed in several ways. The simplest is as a slope expressed as a ratio, 1 in y, meaning that a vehicle rises through one linear unit for every y units it travels horizontally. Thus the angle of inclination a of a ramp to the horizontal is given by $\tan^{-1} 1/y$. Gradients are sometimes expressed as a percentage, say $x\%$, meaning that the ramp rises x units for every 100 units travelled horizontally. For small inclinations a ramp's length is only slightly more than the horizontal distance it spans. Table 6.3 shows the alternative ways of expressing gradients.

Table 6.3. Equivalent slopes, gradients and angles

Slope, 1 in y	Gradient, percentage x	Angle, degrees a
3	33·3	18·4
4	25·0	14·0
5	20·0	11·3
6	16·7	9·5
7	14·3	8·1
8	12·5	7·1
9	11·1	6·3
10	10·0	5·7
11	9·1	5·2
12	8·3	4·8
13	7·7	4·4
14	7·1	4·1
15	6·7	3·8

The slope of a ramp should be shallow enough to be easily negotiated, but at the same time, steep enough to be reasonably short. For example, moving a vehicle from one deck to another 5 m above on a 1 in 14 ramp requires a ramp 70 m long, whilst a 1 in 7 ramp would be only 35 m long but would halve a vehicle's speed [6.1]. Moreover, as slope increases, so does the tractive effort required and also the exhaust emission which introduces problems of ventilation and safety.

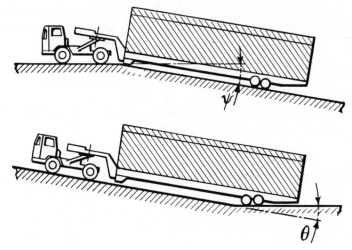

Fig. 6.7. Deck to ramp clearance to avoid grounding of large vehicles. The angles ψ and θ are listed in Table 6.4.

The possibility of vehicles grounding as they move from the deck to the ramp or *vice versa* (Fig. 6.7) must also be considered. Table 6.4 gives the maximum angles permitted for various vehicles to avoid grounding.

It is possible to increase mean inclinations by employing S-shaped, dual slope ramps as illustrated in Fig. 6.8.

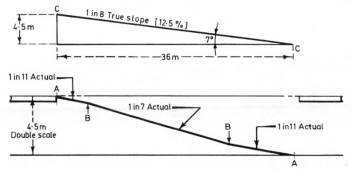

Fig. 6.8 A dual slope S-shaped ramp permits steeper slopes while reducing grounding problems. The lower sketch shows, with vertical scale doubled for clarity, how a steep 1 in 7 centre section with two 1 in 11 transition sections give an overall mean slope of 1 in 8, shown in true scale in the upper sketch. Change of slope is 5 at A and 3° at B, compared with 7° at C.

Table 6.4. Deck/ramp clearance to avoid grounding
(see Fig. 6.7)

Equipment	ψ (ramp/deck) (convex)		θ (deck/ramp) (concave)	
	Slope	Degrees	Slope	Degrees
12.2 m roll trailer	1 in 5·4	10·5	1 in 8·7	6·5
12·2 m road trailer	1 in 4·3	13	1 in 5·7	10

Generally, however, grounding clearance is not the main factor governing ramp slope. Loss of cargo space, vehicle tractive effort and speed, exhaust emission and the danger of uncoupling tractor and trailer have also to be considered. Table 6.5 gives the maximum slopes negotiable by various types of wheeled vehicles, at speeds of 4–5 km/h (1·1–1·4 m/s). It should be noted that steeper gradients can be negotiated by vehicles in good condition but only at reduced speed. Re-starting on the ramp and excessive exhaust emissions can also give rise to difficulties.

Table 6.5. Practical slopes for wheeled vehicles

Equipment	Maximum slope		Preferred slope	
	Slope	Degrees	Slope	Degrees
Car	1 in 5	11·3	1 in 6	9·5
12·2 m trailer or lorry	1 in 6	9·5	1 in 7	8·1
12·2 m roll trailer*	1 in 9	6·3	1 in 10	5·7
Fork lift truck*	1 in 7	8·1	1 in 8	7·1
LUF*	1 in 9	6·3	1 in 10	5·7

* Depending on gross weight; slopes quoted are for weight up to about 60 tonnes.

A virtue of the Ro-Ro configuration is its versatility. Thus it would be unwise for ships to have ramp gradients as steep as 1 in 6 as this would exclude roll trailers, fork lift trucks and LUF systems. The trend is towards a standard ramp slope of 1 in 10. This allows the ready passage of practically all wheeled vehicles in common use, particularly under wet or icy conditions.

In addition, trim must be taken into account. Ships usually trim by the stern, often by as much as 1°. This has the effect of increasing the slope of a 1 in 8 ramp to 1 in 7. Water ballast can be used to correct adverse trim (and heel) at the berth.

Many shipboard ramps have been designed with a slope of 1 in 8 particularly for use on routes where roll trailers are uncommon or can be confined to the main vehicle deck. Alternatively some operators prefer a slope as low as 1 in 14 to allow the rapid passage of heavier vehicles. It should be remembered that vehicle manufacturers' performance figures are invariably given for vehicles in new condition. As a vehicle grows older, its performance falls off and its exhaust emission increases. Some operators confine heavier vehicles to the main vehicle deck to avoid the problems accompanying steep ramps.

The space above the ramp must also be carefully considered. Fig. 6.9 shows the vertical envelope occupied by a 40 ft road trailer and tractor, from which it is clear that deckheads must be carefully arranged to ensure adequate clearance for the container at all stages of its travel.

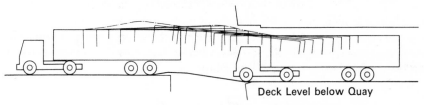

Deck Level below Quay

Vertical "envelopes" determined by 40ft container truck when loading and discharging.

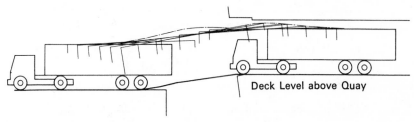

Deck Level above Quay

Fig. 6.9. Vehicle space envelopes for a 40 ft road trailer. When an articulated vehicle ascends or descends a ramp its top describes the arcs shown, so headroom needs to be increased slightly.

6.3.3 Ramp width

Fig. 6.10 shows the ramp widths required for common, wheeled items of Ro-Ro equipment. If ramps are used for two-way traffic, it is usual to allow at least one metre between passing vehicles. Thus a two-lane ramp for roll trailers must have about 7 metres clear width, while fork lift trucks carrying 20 ft containers one way and returning empty require a ramp of 12 m clear width. The full width of the structure is about 10% wider than the clear roadway in axial ramps and about 20% wider in quarter ramps.

40 ft containers are not normally carried at the front of fork lift trucks

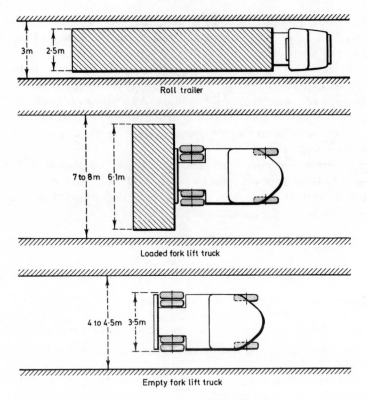

Fig. 6.10. Practical ramp or lane widths for various container-handling vehicles.

aboard ships as happens at container terminals, because their weight and size lead to severe manœuvrability and visibility problems. In practice, 40 ft containers are most popular on the North Atlantic route, where there is ample cellular capacity. Thus the need to transport many large containers in Ro-Ro ships has not yet arisen. When necessary they can be handled by roll trailers. Fig. 6.11 shows the increased width needed to allow a vehicle to change direction while on a ramp or deck. The familiar trapezoidal area of quarter ramps is necessary to provide sufficient space for turning vehicles (see Fig. 6.12). The few examples of curved internal ramps that exist are all fixed.

6.3.4 Ramp length

Ramp length is a function of slope and vertical height. External ramps must accommodate draft changes and tidal variations so that the difference in level between vehicle deck and quay may be within the ship's capability, typically

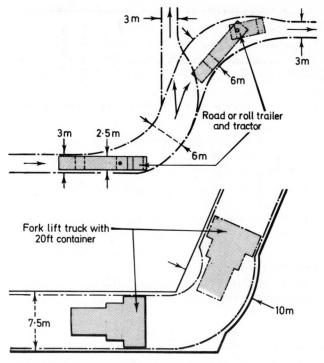

Fig. 6.11. Practical ramp or lane widths for container-handling vehicles changing direction.

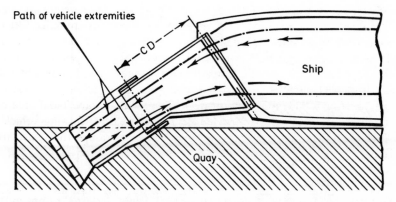

Fig. 6.12. Path of vehicle extremities while traversing a quarter ramp. The critical distance CD is shown, which indicates the minimum length of the first section for it to clear the quay as shown in Fig. 6.13, middle position.

quay 1–2 m above stern door threshold at low water and 4–6 m below at high water with vehicle deck above the quay (see Fig. 6.13).

The design of external ramps must take into account both the arrangement of the ship and the conditions at the ports to which it will call, especially the extreme (spring) tidal range. At a particular port for example, very low tides may occur only about five times a year when the level is likely to be below mean low water springs for about two hours. It is necessary therefore to balance the cost of providing a ramp capable of operating on such occasions against the cost of the rare delays which might otherwise occur.

Many ports now have adjustable bridge ramps so that a ship's stern ramp is used simply to connect with it. Bridge ramps are often positioned at a corner berth as illustrated in Fig. 2.4 and ships using them are equipped with axial ramps.

Axial stern ramps are much shorter than the quarter ramps which rest on the quay. Those on the 18 000 tdw Atlantic Container Lines (ACL) ships are only 15 m long, for example (Fig. 8.13), while the quarter ramp of the 20 000 tdw *Paralla* class is 35 m in length.

The 'critical distance' (CD) is illustrated in Fig. 6.12. It is a function mainly of the greatest permissible quay height above the hinge and the maximum allowable ramp slope.

$$CD = y \times (\text{quay height above ramp hinge} + 0\cdot2 \text{ m clearance})$$

where y is the ramp slope (when expressed as 1 in y). The critical distance must be checked for the low tide/deep draft conditions, as demonstrated in the middle position shown in Fig. 6.13.

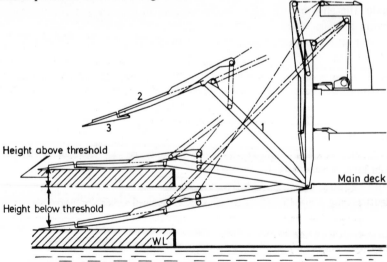

Fig. 6.13. A quarter ramp's opening sequence. This arrangement has a separate stern door. The need for a three-section ramp is illustrated by the middle position where the quay is above the stern door threshold.

Thus the length of the first section of a quarter ramp is the critical distance plus about 1 metre. The second section is long enough for its far end to lie flat on the quay while the third section is entirely on the quay and its area is sufficient to keep the bearing pressure within acceptable limits (see Section 6.3.5).

The length of internal ramps depends on the desired slope, the clear deck height and the depth of the deck structure. Some ships have been constructed with very long internal ramps, e.g. 50 m in the *Paralla*. While internal ramps provide a simple means of moving vehicles from one deck to another, they waste space both above and below. Some space can be recovered by stowing containers or vehicles on and beneath parts of the ramp or by fitting hinged ramps that can be stowed flush with the surrounding deck.

6.3.5 Quay pressure

It has become common practice to limit the maximum bearing pressure that may be exerted on a quay by a ship-to-shore ramp. This limitation has its origins in Australia where a number of conventional quays, which were to be used as Ro-Ro berths, were not strong enough to support the concentrated load of the end of the ramp and a vehicle traversing it.

The designed maximum quay pressure exerted by a ramp depends on the permissible quay loadings at the ports where the ship is expected to call. To achieve low bearing pressures, the third ramp section which rests on the quay must be large in area. This requirement adds to the ramp's initial cost. Thus when fully laden fork lift trucks can exert quay loadings of 50–60 tonnes per axle, the ship operator could be wasting money by demanding maximum ramp pressures of around 2 tonnes/m^2, as is common.

Actuating cables are attached to the first ramp section of quarter and slewing ramps and take some of the weight. Where axial ramps rest on the quay their weight is supported both by the quay and the shipboard hinge.

In large Ro-Ro vessels, a stream of fork lift trucks crossing a quarter ramp with 20 ft containers can cause the ship to heel 2–3°. With a beam of 30–35 m, the difference in level between ship and quay would change by 0·5–1·0 m. This must be taken up by the ramp, and can be done by using fast response winches which adjust automatically to the passage of each vehicle. On large ramps this is an expensive solution as it uses a great deal of power and requires complex control systems [6.1]. An alternative is to install a buttressing hinge between the first and second ramp sections. In effect, this is a hinge that can be adjusted to any desired angle and then locked. A proportion of the ramp weight is then taken by the actuating cables (which are also used to hoist the ramp into its stowed position). The remaining proportion of the ramp weight and that of the vehicles on it ensures that the ramp remains firmly on the quay at all times.

This arrangement ensures a firmer connection between the ship and shore and results in reduced angles of heel. The vertical distance between the ramp and quay is measured by sensors, attached to the ramp, that actuate the buttressing mechanism to compensate for changes in tide level and draft, but not for the passage of each vehicle. This allows the response capability of the winches to be reduced.

6.3.6 Construction

Ramps consist of longitudinal beams plated over to provide a vehicle roadway. Two massive beams along each side of the ramp provide much of the strength, with additional strength contributed by closely spaced longitudinals between them, as illustrated in Fig. 6.26. Transverse members stiffen the structure locally. Alternatively a grillage of equal-depth longi-

Fig. 6.14. The second and third sections of long ramps are split as shown to provide transverse flexibility, allowing the ship to heel while the third section of the ramp remains firmly on the quay. A non-slip surface is provided by welded strips. This 22 m long ramp is on the *Strider* Class vessel whose profile is shown in Fig. 8.6.

tudinal and transverse stiffeners can be used, as shown in Fig. 6.21. Additional small steel strips are welded to the roadway, often in a herring-bone pattern, to provide an anti-slip surface (Figs. 6.14 and 6.20).

In order to design the most economical ramp, access equipment manu-facturers must know the gross vehicle weights, axle loads and spacing, and the wheel loads that are expected. Any of these can be the critical design parameter. Such parameters help to determine the scantlings of the ramp members because the deflections of plating between stiffeners, and the strength of the stiffeners, are as important as the overall strength of the ramp.

Ship operators are often forced to carry heavy vehicles with different axle configurations and weights from those for which their stern ramps were

Fig. 6.15. Folded stern ramp. Note that by folding the ramp when at sea, its centre of gravity is lowered and its effect on the air stream over the stern of the vessel is lessened.

Fig. 6.16. The 5700 tdw freight Ro-Ro *Orion* berthed at Hull. Since the dock is enclosed, her twin stern ramps can rest directly on a shore ramp. Two roll trailers loaded with timber are being towed by terminal tractors.

initially designed. A chart, similar to the one shown in Fig. 9.12, can be drawn up for an axial ramp to indicate whether or not it can support a given load.

The ramp attachment points on the ship must be strong enough for it to function correctly. The manufacturer's responsibility extends to the ramp alone; the necessary supporting structure for it must be provided by the ship designer.

6.3.7 Heeling

It is necessary for the end of the ramp to remain flat on the quay regardless of any heeling, trimming or ranging of the ship. Most operators expect their ramps to accommodate heels of up to 5° towards or away from the quay. An example of how this may be done by splitting the second ramp section longitudinally is illustrated in Fig. 6.14. Here the third section, which rests on the quay, is also split.

This arrangement may not be suitable when excessively heavy loads are transmitted across the hinges. In such cases the steelwork is designed to be inherently flexible with the width of each portion not exceeding 95% of its length [6.2] and the final transition to the quay is provided by a number of finger flaps, each approximately 0·5–1·0 m wide, as shown in Figs. 6.20 and 6.28.

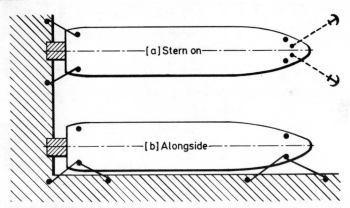

Fig. 6.17. Alternative berthing arrangements for vessels equipped with stern ramps, where quay heights do not necessitate a bridge ramp.

6.3.8 Stowage

The simplest method of securing a ramp for sea is to raise it as a single section into the vertical position with winches and to cleat it against the ship's structure. This method is commonly employed for axial ramps, although quarter ramps, up to about 20 m long, can also be stowed in this way. The high vertical centre of gravity of longer ramps can adversely affect a ship's stability, while their considerable windage area can impair its steering and cause exhaust fumes to be sucked into its ventilation system. Thus, the second and third sections are usually folded down as shown in Fig. 6.15.

6.3.9 Types of external ramps

6.3.9.1 *Axial ramps*

An axial ramp is the simplest bridge between ship and shore, but it is somewhat limited in its applications. Fig. 6.16 shows twin axial ramps lying directly on the quay.

Axial ramps are ideally suited for Ro-Ro vessels intended to serve specific ports where berthing stern-on or in a corner is possible, as illustrated in Fig. 6.17. The arrangment shown in Fig. 6.17(a) requires calm weather conditions and sheltered waters. For ships with short ramps, a small tidal range is also necessary. Stern-on berthing is not always practicable, especially for large ships. Thus, the more usual arrangement is to berth in a corner as shown in Fig. 6.17(b). Ships with axial ramps are usually employed on established routes where bridge ramps that can accommodate most of the tidal variations, are avialable.

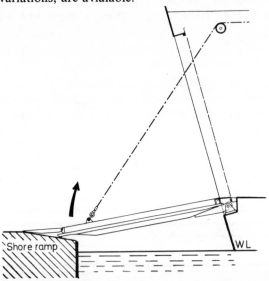

Fig. 6.18. A short axial ramp which also forms the stern door in its closed position. This ramp is wire-operated.

An axial stern ramp that doubles as a stern door when raised is shown in Fig. 6.18, and an alternative arrangement, where the ship has a long axial ramp and a separate weathertight door, is shown in Fig. 6.19.

When lowered, ramps can either rest on a bridge ramp or on the quay as shown in Fig. 6.18. The weight of the ramp and vehicles is then shared by the quay and the ship. Alternatively, most of the weight can be supported by chains as shown in Fig. 6.19 to limit the quay bearing loads (see Section 6.3.5). Supporting chains can be adjusted to allow for tidal movement and changes in the vessel's draft and trim. Axial ramps can be operated by hydraulic rams, chains or, most often, by hydraulic winches and wires.

Axial ramps sometimes provide access to two decks. Occasionally, one side of the ramp leads to the main vehicle deck while the other connects with an internal ramp. In other cases two short axial ramps are tiered and both can connect the ship to a bridge ramp. This arrangement does not usually allow simultaneous loading of two decks however, unless a two-tier bridge ramp is used.

A typical axial ramp is 10–15 m long, 5–8 m wide and weighs about 0·4 tonnes/m², depending on dimensions and loading (see Section 6.9).

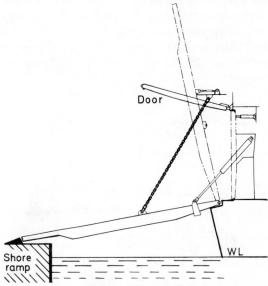

Fig. 6.19. A longer axial ramp which has a separate stern door. This ramp can be supported by a non-adjustable chain to reduce the pressure on the quay. The ramp is raised by a hydraulic cylinder.

6.3.9.2 *Quarter ramps*

Quarter ramps are often installed in vessels intended to trade to ports where only conventional quays exist, or in vessels that have not been designed for a specific route and thus require a 'go-anywhere' capability. Fig. 6.20 shows a quarter ramp of the slewing type (see Section 6.3.9.3) and demonstrates its most important characteristic: its ability to turn a conventional general cargo berth into one suitable for Ro-Ro operations, provided that it is clear of bollards and other obstructions.

Quarter ramps have been fitted to Ro-Ro vessels of sizes ranging upwards from the 4200 tdw *Akademik Tupolyev* which has a three-section ramp 22m long, to ships of over 30 000 tdw with 50 m ramps. Experience has shown, however, that quarter ramps are not very practical in small ships (see Section 8.3.2.2).

Fig. 6.20. Angled quarter ramps or slewing ramps can give access to conventional quays. The 35 m long 7 m wide slewing ramp on the 15 000 tdw deep sea Ro-Ro *Reichenfels* is shown together with the separate stern door. Only the third section with its finger flaps rests on the quay. The two towers which support part of the ramp weight and control its slewing are clearly seen.

Fig. 6.13 shows, in diagrammatic form, the opening sequence of a large three-section quarter ramp, which stows folded to lower its centre of gravity, reduce its windage, and improve visibility over the stern. The first of the three hinged sections of a quarter ramp is usually supported by cables and hydraulic winches.

Quarter ramps are usually stowed with a slight inboard inclination, from which position they are pushed by hydraulic rams when being lowered. After this initial nudge, the main actuating cables take over, lowering the ramp

Fig. 6.21. A two-sided ramp, which can be used either as an axial ramp or as a quarter ramp if a separate ramp section is attached to the angled part of the axial section by means of the crane to dock with either side alongside the quay. The grillage construction is clearly seen.

under gravity. A secondary spreading arrangement, consisting of a pair of hydraulic rams or another winch, opens out the three-ramp sections to the desired position on the quay. Lowering a quarter ramp into its normal working position usually takes between 10 and 20 minutes, while a simple axial ramp takes only a minute or two. A separate stern door is necessary in ships with quarter ramps to provide weathertightness.

Fig. 6.22. Twin quarter ramps on a ship converted to Ro-Ro operation, which allow it to berth alongside a conventional quay either to port or to starboard.

Fig. 6.23. A Japanese ferry fitted with both an axial and a quarter ramp. This arrangement allows the vessel to utilize a Ro-Ro berth when available or alternatively a conventional berth. The ramps are not used simultaneously.

A variant of the conventional quarter ramp is shown in Fig. 6.21. This is basically an axial ramp, but it is angled at both outer sides, to which a second portable section can be attached by means of a ship-board crane, if required as a quarter ramp. The extension is stowed on deck when not in use. Ships with ramps of this type can usually berth either port or starboard side to the quay. This arrangement is, however, not widely used.

Another variant is shown in Fig. 6.22 where a ship can berth either side to the quay, by having twin quarter ramps. Some ships have a quarter ramp and an axial stern ramp or door. Notable examples include the *Paralla* and the Japanese ferry *Orion*, illustrated in Fig. 6.23. It is understood, however, that the *Parallas* rarely use their axial door in the Australian trade.

It is also possible to fit an angled ramp at the forward shoulder of a ship. This arrangement, providing drive-through facilities, is common in Japanese inter-island ferries, such as the *Golden Okinawa* illustrated in Fig. 6.24. Quarter ramps are usually fitted on the starboard side (exceptions include some Australian Ro-Ro vessels), so that ships have to berth starboard side to.

Fig. 6.24. A forward mounted angled ramp on a Japanese ferry. When fitted to passenger/vehicle ferries equipped for one way traffic flow, this ramp takes the place of a conventional bow ramp and allows the vessel to use an ordinary quay.

Typical quarter ramps, as shown in Fig. 6.61, weigh about 0·6 tonnes/m². The 68-tonne quarter ramp on the *Akademik Tupolyev* has a 92 kW hydraulic installation which raises or lowers it in 9 minutes, a typical power requirement for this size of ramp. Larger stern ramps require larger power units – 200 kW or more.

6.3.9.3 *Slewing ramps*

It sometimes happens that a Ro-Ro vessel has to berth with its quarter ramp away from the quay. If this is likely to happen often, the attendant delays might make a slewing ramp attractive (rather than the more cumbersome alternatives outlined in the previous section). A typical slewing ramp is shown in Figs. 6.20 and 6.25.

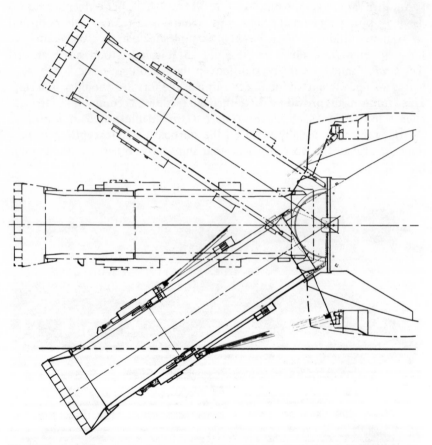

Fig. 6.25. A slewing ramp of the type illustrated in Fig. 6.20. It can be positioned to any angle between 33° port and starboard of the centre-line, so that complete flexibility is obtained using conventional quays.

Slewing ramps have three sections hinged together as in a quarter ramp. The outboard end of the first section is suspended by cables from hydraulic winches, while its inboard end is hinged to permit vertical movement and supported by a king pin to allow horizontal rotation. Various slewing arrangements are used. In some cases the king pin rotates as the ramp is turned by a jigger winch, sliding on a semi-circular, lubricated bearing

Fig. 6.26. The slewing ramp on the Ro-Ro *Reichenfels* during the opening process. The ramp stows vertically in the centre position and is unfolded through about 30° before the ramp is slewed and the second and third sections unfolded. The construction using deep longitudinal girders can be seen, as well as wires controlling the various sections.

surface. In other cases, the ramp rotates about a king pin, rather like a large derrick which is slewed by adjusting its topping lifts. Since the cable winches are anchored to the ship's structure, the ramp can be slewed by shortening the cable on the side to which it is to move (and lengthening the other one). Rack and pinion and hydraulic mechanisms are also used for slewing.

Slewing ramps stow axially and must be unfolded and lowered through approximately 30° before they can be slewed as shown in Fig. 6.26. Slewing ramps can sometimes serve as short axial ramps when the second and third sections can remain folded and only the first section is used.

From its stowed position, inclined slightly inboard, the ramp is pushed away from its buffers by a pair of hydraulic rams, and lowering then

Fig. 6.27. A bow ramp fitted in a small Ro-Ro passenger/vehicle ferry permits the drive-through facility useful for rapid handling of road-going vehicles or tourist cars. The opening is closed by a bow visor, while a separate inner water-tight door is provided.

proceeds under gravity. Once the ramp has reached an inclination of about 30° to the vertical, the ramp is slewed and the second and third sections are opened out by a separate spreading winch. After this, final adjustment is made as required. A separate weathertight stern door must be fitted in ships with slewing ramps.

Slewing ramps are usually designed to slew about 30°–40° off the centre-line. Vehicle turning space has to be provided inside the ship, in contrast to vessels with quarter ramps having trapezoidal sections which offer sufficient room for vehicles to change direction. A simpler form of semi-slewing ramp has been designed, capable of slewing right aft or to one side only, as illustrated in Fig. 8.3.

6.3.9.4 *Bow ramps*

These are usually of the axial type, and link the ship to the quay through a bow opening having a visor or shell doors, as shown in Figs. 6.27 and 6.28.

In construction and design, bow ramps are very similar to axial stern ramps. When raised, they form an extension to the collision bulkhead as required by Classification Society rules (see Section 4.3.4). This is illustrated

Fig. 6.28. An alternative closure for a bow ramp is by twin shell doors, whose construction is clearly seen.

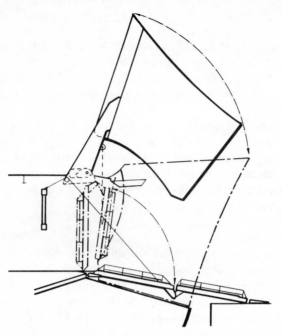

Fig. 6.29. Bow visor and ramp arrangement with the folding ramp forming in the stowed position the inner door required by regulation.

Fig. 6.30. Prow door/ramp installation used on small ferries in sheltered waters. The configuration is similar to that of a landing craft, with the ramp leading directly onto the open vehicle deck.

in Fig. 6.29. Longer ramps may have to be folding or telescopic to stow within the hull.

A variant of the bow ramp is the prow door/ramp used in small ferries operating in sheltered waters, shown in Fig. 6.30. This forms the only means of closing the bow opening, as, for instance, in small landing craft. A similar arrangement can be used in small cargo vessels, where vehicles can be driven onto the tops of the hatch covers.

Bow ramps are usually 10–15 m long and 3–4 m wide, weighing about 0·4 tonnes/m².

Fig. 6.31. A fixed ramp leading from the main vehicle deck to the lower deck with a hinged articulated hydraulically operated cover, shown here in the open position. In the closed position the cover has additional cargo stowed on it.

6.3.10 Types of internal ramps

6.3.10.1 *Fixed ramps*

Fixed internal ramps provide access to decks above or below the main vehicle deck served by the external ramp. They are usually found in large ships where wasted space is a less serious problem than in small ships. Considerations governing dimensions are discussed in Section 8.3.2.3. Their construction is usually simple and they are designed with the slopes suggested in Section 6.3.2. In vessels in which fixed ramps provide access to spaces below the freeboard deck, the opening in the freeboard deck should be provided with some means of closing. A side- or end-hinged hydraulic cover is usually employed for this purpose as illustrated in Fig. 6.31.

Internal ramps are provided with a non-slip surface, which may be welded steel strips or expanded metal, or alternatively a non-slip coating.

6.3.10.2 *Movable ramps*

Movable or stowing ramps are often used to provide access to car decks. Once the car deck is full, additional vehicles can be parked on the ramp which can be raised into the horizontal position for this purpose, as shown in Fig. 6.32. When not in use, movable ramps can be stowed under the deckhead to provide as much headroom as possible.

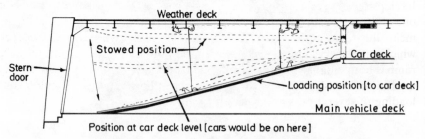

Fig. 6.32. A movable ramp giving access to a car deck. This example is S-shaped to give easier access to and from the decks while allowing a steep gradient. When not in use it can be hoisted up against the deckhead, thereby increasing the headroom below. Vehicles may also be carried while it is raised to the car deck level.

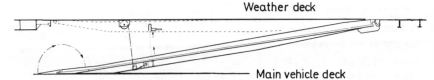

Fig. 6.33. Combined hatch cover and ramp. This ramp forms a hatch cover in its upper stowed position. When used to give access from the main deck to one above it can be hoisted into its stowed position with a vehicle lashed to it, and further vehicles are then stowed below, thus wasting little space.

6.3.10.3 *Hatch cover/ramps*

Hinged ramps can be used both to provide access to spaces below the freeboard deck and to provide a means of closing the opening in the freeboard deck at sea. A hinged hatch cover/ramp is illustrated in Fig. 6.33. An alternative arrangement, which allows vehicles to be parked on the ramp before it is lowered into its stowed position, is shown in Fig. 6.34. This avoids the waste of space usually incurred beneath fixed ramps.

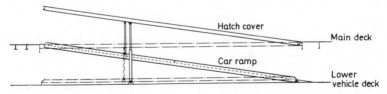

Fig. 6.34. Hatch cover and vehicle ramp combination. The ramp and cover move simultaneously and if used to give access from the main deck to one below can have vehicles lashed to both the cover and the ramp.

6.3.10.4 *Multi-purpose ramps*

In ships having several vehicle decks, less space is wasted beneath ramps if one of them is of the multi-purpose type, as shown in Fig. 6.35. The arrangement illustrated allows access to four decks, from the main vehicle deck which is itself served by a stern ramp. In the stowed position a multi-purpose ramp is horizontal and provides additional deck area. Such ramps are mostly used in smaller ships, where there are fewer opportunities for simultaneous loading of several decks than in larger ships.

6.4 Bow openings

Bow openings allow access to cargo spaces through the forward end of the ship as mentioned in Section 6.3.9.4. Two types of closing are commonly fitted, bow visors and bow (shell) doors.

6.4.1 Bow visors

A bow visor is illustrated in Fig. 6.27. It is a portion of the ship's bow, hinged at weather deck level and usually provided with gaskets to restrict the entry of water when closed. Visors are usually opened hydraulically as shown, although a few are operated by winches. Their weight is generally within the range 30–50 tonnes. They are secured by manual or hydraulically-operated cleats.

6.4.1.1 *Construction*

Externally, the construction of a bow visor appears like a normal ship's bow, but it is stiffened internally and has strong pivot arms welded to the deck as shown in Fig. 6.29. To obtain an accurate fit, the bow structure is usually built as part of the ship and then cut and refitted as a visor. Bow visors are sometimes fitted to ships that have been converted into Ro-Ro vessels and when this is done, the original bow section can sometimes be removed, strengthened, fitted with the actuating mechanisms and seals, and then replaced (see Section 10.8.2.5).

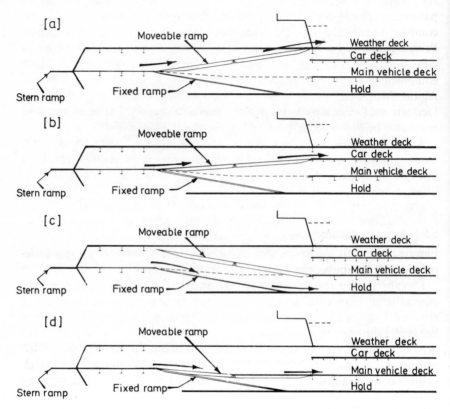

Fig. 6.35. Multi-purpose ramp. When a vessel has many vehicle decks, this type of ramp can be used to give access from the main vehicle deck to several other decks, wasting a minimum of cargo space.

6.4.2 Bow doors

Bow (shell) doors are located in the plating around the stem. They are usually fitted in pairs and hinged to swing outwards as shown in Fig. 6.28. Their operation, sealing and securing arrangements are similar to those of bow visors. They must be carefully designed to ensure that they do not foul the ship or bridge ramp when being opened or closed.

6.4.2.1 *Construction*

Like bow visors, bow doors are similar in external appearance to a conventional ship's bow. Internally they are heavily stiffened, however, as is the supporting hull structure.

6.4.3 Use of bow access

As pointed out in Section 6.1.3.2, bow access is mainly used in short sea passenger/vehicle ferries where traffic flow is one-way, and drivers accompany their vehicles for the voyage and drive them off at their destination. Private cars and lorries are usually carried, the largest of which requires a clear opening about 3 m wide and 4·5 m high. The design of bow access equipment is fairly complicated because of the shape of the hull in this region. Moreover, the internal bow door/ramp must be long enough to facilitate use, while stowing within the available space. A second inner door must also be provided as discussed in Section 4.3.3.

Ships with bow access usually have their windlass, chain locker and hawse pipes located further aft than usual so that they are well clear of the visor or shell doors.

6.5 Elevators

6.5.1 General considerations

Elevators (or lifts) are used as alternatives to ramps for transferring vehicles or other cargo above and below the main vehicle deck.

Elevators require considerably less internal ship space than ramps, especially in high deck spaces. Cargo can be stowed on some types of elevator, thereby allowing better space utilization than ramps. Moreover it has been estimated that a vehicle ascending a ramp emits 3–4 times as much exhaust gas as a vehicle idling on an elevator [6.2]. This means that vessels fitted with elevators can have lower-capacity ventilation systems than vessels equipped with ramps. Such savings can be set against the higher first cost of elevators.

Disadvantages of elevators include their interrupted cycle of operation, unlike ramps which can be used continuously, and the possibility of malfunction delaying cargo operations.

6.5.2 Types of elevator

6.5.2.1 *Wire-operated platform*

The wire-operated platform is the simplest form of elevator and is illustrated in Fig. 6.36. It consists of a rectangular vehicle platform guided vertically at its four corners and operated by a hydraulic jigger winch as shown. In its simplest form it stows at its lower level in a recess let into the vehicle deck or tank top. In this position it can support a vehicle securely lashed to the platform, but a cover is needed to close the upper opening. Often a hydraulically operated, side-hinged cover is used for this purpose.

In an alternative arrangement, the elevator acts as a hatch cover when it is secured at the upper limit of its travel. The joint between the platform and deck can be made weathertight if necessary. This type of elevator can also be supplied with hinged guides which do not obstruct vehicle decks when not in use.

Wire-operated elevators can lift vehicles weighing up to 50 tonnes at an average linear speed of 5–6 m/min. A typical overall cycle time to load a vehicle, lower, discharge and rise again would be about 3 minutes. Their platform dimensions are usually around 16 m by 3·5 m with a working height of 5–6 m. The whole installation may weigh about 30 tonnes, excluding any additional structure round the deck opening, and can be arranged to serve several decks.

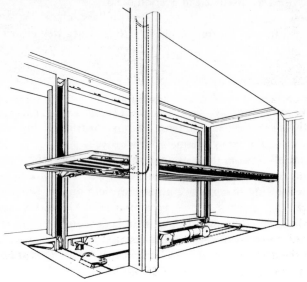

Fig. 6.36. Wire-operated elevator, used in conjunction with a hatch cover above.

6.5.2.2 *Chain-operated elevator*

Similar in principle to the wire-operated type, this elevator uses heavy roller chains to lift its platform, as illustrated in Fig. 6.37. The chains run over sprockets on the platform and are attached to horizontal hydraulic rams which raise or lower the elevator as they extend or retract. Sprockets at opposite sides of the platform are connected by axles to ensure that the elevator remains level during operation.

One advantage of this type of elevator is that the sprocket diameter is much smaller than the sheave diameter of the equivalent wire-operated elevator. The elevator, like other types, can be guided at only one edge. If

Fig. 6.37. Chain-operated elevator. The platform forms the hatch cover in its upper position. Note the end flaps for access to the lowered platform which does not recess into the deck below.

it is located at the ship's side, the guide can be arranged to avoid obstructing the vehicle deck as shown in the illustration.

Like wire-operated elevators this type can provide a weathertight means of closing the deck when stowed at the upper limit of its travel. Hydraulic oil is fed to the cylinders on the platform by flexible hose, from a supply point on the deckhead above. Chain-operated elevators do not usually have a lower recess in the deck. They are therefore provided with flaps at each end of the platform so that vehicles may be driven onto them.

This type of elevator is best used to give access from the main vehicle deck to upper decks, stowing at the upper limit of its travel. It can support a vehicle when stowed and thus wastes no space on the upper deck. Moreover, since it does not obstruct the lower deck, there is no space wasted beneath it either.

A variant of the four-chain system has platform supports cantilevered from vertical guides and uses only two chains or wires. This leaves one edge of the platform clear and thus allows vehicle access over three sides. Such elevators are sometimes referred to as L-type lifts, as indicated in Fig. 8.12.

6.5.2.3 *Scissor lifts*
Scissor lifts (elevators) have a platform supported by a system of levers and hydraulic rams as illustrated in Fig. 6.38. They do not need guides and when

Fig. 6.38. Scissor lift. Usually installed to give access to the tank top, the scissor lift is normally used in conjunction with a hatch cover at the main deck, as it stows recessed into the double bottom.

stowed are recessed into the double bottom as shown. Often, small vertical hydraulic rams are fitted as well as the main rams to provide additional upthrust, when the platform is at its lowest position and the leverage of the main rams is small.

Scissor lifts are usually located between the main vehicle deck and lower decks, stowing in a recess at the lower level. A hatch cover is therefore needed to close the upper opening. No cargo space need be lost with scissor lifts.

A typical installation might have a lifting capacity of 30 tonnes with a platform measuring 13 m by 3 m and operate through a vertical distance of 4–5 m. Such an installation would weigh about 25 tonnes. Operating speed is usually around 5 m/min and the lower deck recess is between 0·6 and 0·8 m in depth.

6.5.2.4 *Double platform elevator*
Double platform or two-level elevators are used on ships with at least three vehicle decks (including the weatherdeck) for transferring vehicles rapidly

Fig. 6.39. Double platform elevator. Used to serve three or more decks, this arrangement is cheaper than two single elevators, but careful traffic control is needed to avoid delays.

between decks. They are usually chain-operated as shown in Fig. 6.39, but may also use direct cylinders.

For vessels having equal deck heights above and below the main vehicle deck, the loading sequence is as follows;

   (i)   load upper platform at main vehicle deck;
   (ii) raise to upper deck level;
  (iii) unload upper platform;
  (iv) load lower platform at main deck level;
   (v) lower to tank top level;
  (vi) unload lower platform and return to (i).

In vessels having unequal deck heights, an extra step (iiia) is needed to bring the lower platform to the main vehicle deck level.

Although there is usually never more than one vehicle on a double platform elevator at any given moment, its use does save time by omitting empty movements. Moreover, it is cheaper to install than two one-level elevators (see Fig. 8.12). For maximum throughput, however, this system does require that upper and lower decks are worked simultaneously.

Double platform elevators are usually stowed in their lower position with a vehicle on both platforms. A hinged hatch cover closes the opening in the upper deck. Only two vertical guides are necessary, which are usually located against the ship's side.

Typical platform dimensions for a 40-tonne capacity elevator are 19 m by 4 m, with an operating speed of 4–5 m/min. The total weight of such an installation is between 50 and 60 tonnes.

6.5.3 Powering systems

As already mentioned, elevators are raised and lowered either by jigger winches with wires or by chains and hydraulic rams or by levers and hydraulic rams. These options are confined to elevators of less than about 65 tonnes capacity. For higher capacities winches are necessary [6.3].

A typical 50-tonne S.W.L. (safe working load) elevator requires two 75 kW pump units. Power units for elevators must move the platform smoothly, starting slowly, speeding up and then slowing down towards the end of its travel. This is achieved by variable delivery pumps.

Hydraulic rams and chains are becoming the most popular drive mechanisms. Moreover the use of chains, with a safety factor of around 2·7, is acceptable to various authorities who would otherwise require double wires, larger sheaves and other safety measures with jigger winches.

6.5.4 Control systems

Elevators are designed to be contiguous with the decks at both ends of their travel, where they are held in position by locking devices. Loads imposed on the platform as vehicles move onto it are thus supported by the locking devices rather than the power unit, and the platform will not fall in the event of a failure in the lifting mechanism.

When the lifting mechanism is actuated, its control system first unlocks the platform. Sensors in the elevator guides feed the position of the platform back to the control unit so that it slows down over the last 150 mm or so of its travel. An inclinometer provides an audible warning to the operator if the ship's heel exceeds a preset value, usually about 2°. Elevators cannot operate against a heel of 5° or more. Fail-safe devices lock the platform automatically if a major lifting mechanism failure, such as a wire breaking, occurs.

To ensure the safety of personnel, fencing must be rigged around the edges of elevator deck openings when the platform is in use at another deck. Some elevators stop automatically if cargo is protruding beyond the edge of the platform.

6.5.5 Sealing

Where elevators are secured at the upper limit of their travel to act as weathertight covers, gaskets are inserted around the openings under the deck. In normal operation, platforms are designed to stop just short of these gaskets to reduce wear from continual contact, but the securing operation raises the platform the last 10 or 15 mm necessary to compress its rubber gasket. The platform is cleated in this position ready for sea.

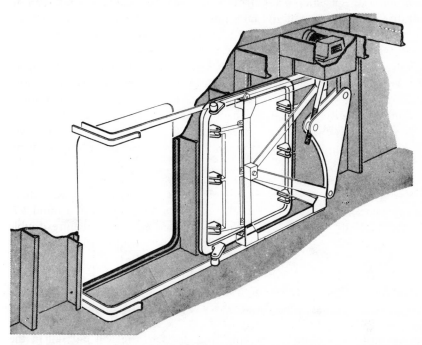

Fig. 6.40. Inward-opening side door. Usually fairly small, this type of door is used for ship's provisions and stores of to give access to cargo spaces for small unitized loads.

6.5.6 Use of elevators in ships

Elevators waste less space in a ship and give rise to less exhaust emission from vehicles, than internal ramps. But fixed ramps cannot easily fail and movable ramps are less complicated than elevators. They are also continuously available.

Moreover, the locations of elevators within a ship must be carefully chosen. If they are too close together when serving different (as opposed to the same) decks the manœuvring space available for vehicle streams is likely to be restricted. It has been estimated that approximately 150 m² of manœuvring space per elevator is required by fork lift trucks [6.4]. Section

8.3.2.3 discusses further the question of elevators versus ramps, for ships of different sizes.

6.6 Side doors

6.6.1 General

Side doors (or side ports) are found in both general cargo and Ro-Ro vessels. Their purpose is usually to augment the main cargo accesses, but not always. Side doors can best be employed in ships whose draft remains fairly constant, on routes where tidal variations are small. Where side ports are well above the waterline, doors are sometimes not fitted, as in large U.S. trailer-ships, as long as all openings in the freeboard deck are made watertight.

A variety of different side doors are available. The individual design

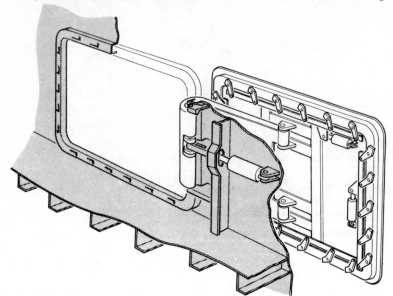

Fig. 6.41. Side-swinging door. It can provide an extremely large opening if required.

depends largely on the type of cargo carried, the handling method employed and the distance between quay and ship. Side doors are particularly suitable for palletized cargo. Some ships have been built to handle their cargo through side doors exclusively. The 7000 tdw *Zaida* (renamed *Vendee*), a refrigerated pallet carrier, is an example of one such. But, with the exception of the fruit and a few other specialized trades, pallets are less widely used than containers.

Fork lift trucks are normally used for handling pallets, bales, drums and similar cargo. One truck operates on the quay, transporting the cargo from a shed to a platform at the side port, whence it is transferred to its stowage position by other fork lift trucks, conveyors or elevators.

6.6.2 Types of side doors

6.6.2.1 *Inward opening doors*

One simple type of side door, illustrated in Fig. 6.40, opens inwards and slides longitudinally to lie along the ship's side. Driven by an electric motor connected to a screw or endless chain system, this type is self-closing. A guide forces the door against its frame and it is then cleated manually.

Such doors are usually fairly small and are used mainly for ship's stores and small cargo pallets.

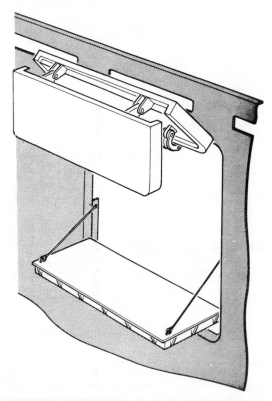

Fig. 6.42. Upward-folding door. This type is usually supplied with a loading platform as shown.

6.6.2.2 *Side swinging doors*

Illustrated in Fig. 6.41, the side swinging door is hydraulically operated and secured. It opens outwards and stows along the ship's hull. This type of door can be used for larger openings than inward sliding doors. It is sometimes installed in pairs, one each side of the access port, and can be provided with an adjustable loading platform similar to the one shown in Fig. 6.42, where cargo can be set down.

6.6.2.3 *Upward folding door*

This door is hinged at its upper edge and also at its mid-length so that it folds upwards to stow against the hull above the opening as shown in Fig. 6.42. A loading platform folds out from the bottom of the opening to lie horizontally above the quay. Both the door and loading platform are hydraulically operated, the latter with wires from a jigger winch.

6.6.2.4 *Side door/hatch cover*

This is a two-section hatch which folds up and back hydraulically, as shown

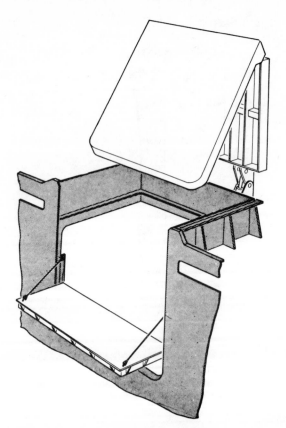

Fig. 6.43. Side door/hatch cover. Pioneered by Fred Olsen. Used when the units to be handled are high, or when some degree of vertical access required.

in Fig. 6.43. Like other types it has an adjustable loading platform that swings out through the opening for cargo from fork lift trucks ashore. These covers are usually adopted in ships calling at ports with relatively high quays. They are used as illustrated in Fig. 6.46 and can also be used for limited vertical access.

6.6.2.5 *Side port conveyor and elevator*

Similar in construction and use to the side door/hatch cover, this system has an additional loading platform inboard of the outer one normally fitted, as shown in Fig. 6.44. This is to facilitate the transfer of cargo to lower decks.

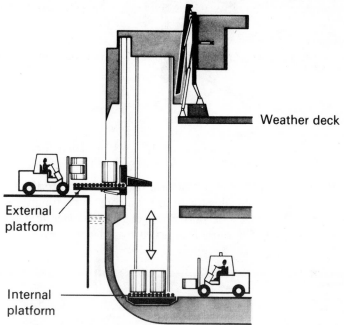

Fig. 6.44. Side port conveyor and elevator in use in a multi-deck operation.

In use, the inboard elevator platform is raised to the level of the outer platform. When it is at the correct height, sensors set conveyor rollers in motion to move cargo from the outboard to the inboard platform. This then descends to a lower level, where the cargo is retrieved by a fork lift truck and stowed. The system is hydraulically operated and completely automatic.

6.6.2.6 *Side door/ramp for fork lift trucks*

This type of side door fulfils the same function as the side port conveyor and elevator system described in the previous section. The only difference is that the external loading platform is in the form of a ramp (Fig. 6.45) up which fork lift trucks can drive to deposit their loads within the ship as shown

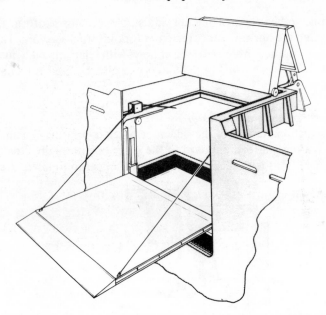

Fig. 6.45. Side door and ramp for fork lift truck operation. Structural compensation is required at the deck and side shell for the openings.

in Fig. 6.46. An inboard elevator platform takes the cargo to the required level. This arrangement obviates the need for a conveyor with an elaborate control system. Like other side port arrangements, the side door/ramp is hydraulically operated. The ramp is adjusted by a jigger winch.

Typical loading possibilities using side door/ramp.

Fig. 6.46. Side door and ramp in typical operation, with fork lift trucks handling pallets both on the quay and in the ship.

6.6.2.7 *Side door/ramp for Ro-Ro operation*

Side ramps are similar in principle to the axial ramps described in Section 6.3.9.1 but are in two or three sections as shown in Figs. 6.47 and 6.48 to limit the maximum slope. This allows the ramp to fold to form a weathertight door when raised. Side ramps are used mainly in car carriers and vessels that use trailers or fork lift trucks to stow directly.

Fig. 6.47. A side door/ramp 6·4 m wide by 6·2 m high in the 22 000 tdw *Seaspeed Asia*. These dimensions permit the passage of roll trailers loaded with containers two-high, or large fork lift trucks.

The first ramp section in Fig. 6.48 is lowered by wires and a jigger winch. The second section is unfolded by hydraulic cylinders and the third and smallest section is unfolded manually. A typical door opening is about 2·5–4 m square, the larger sizes allowing the passage of fork lift trucks as well as motor cars.

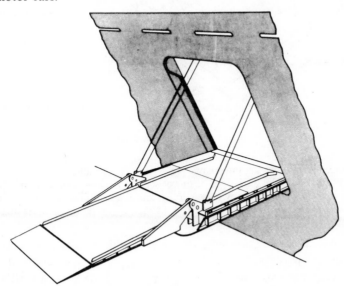

Fig. 6.48. Side door/ramp for Ro-Ro operation where a fairly long three-section ramp is required to accommodate appreciable tidal and draft variations.

6.6.2.8 *Emergency side door for Ro-Ro operation*

As Ro-Ro vessels become more and more complicated and cargo handling arrangements depend increasingly on the satisfactory working of large stern ramps, some ship operators have taken steps to provide alternatives as insurance against serious malfunctions of the vital stern ramp. A simple and cheap side door, about 4 m square, is fitted in some cases, while in others side ports designed to receive a shore-based ramp are installed. A door suitable for emergency or occasional use is shown in Fig. 6.49. It is massively constructed to allow the passage of large, heavy equipment and its platform can be equipped with a king pin and a slewing arrangement to alleviate difficulties that may be experienced when units enter or leave the hull. Such side doors are larger than the doors considered so far, with openings measuring approximately 4·5 m square and doors weighing about 5 tonnes.

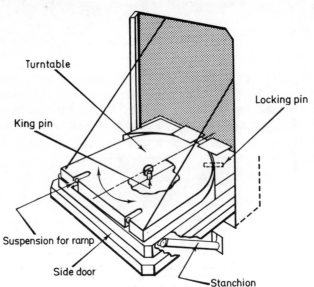

Fig. 6.49. Emergency side door for Ro-Ro operation. Fitted to ships with complex stern door arrangements, this door is designed to allow traffic movement should the stern door be out of service. A shore-based ramp is fitted to the suspension points shown by a crane, and the turntable can be rotated to reduce the direction change required by the vehicle on entering or leaving the ship.

6.6.3 Use of side ports in ships

As already discussed, side ports or doors are often used to augment whatever other cargo handling arrangements a ship may have. Although pallet dimensions are not standardized, many in current use measure about 1 m by

1·2 m and can be loaded up to 1·8 metres high. Thus a side port must be at least 2 to 2·5 m square. Where tidal ranges and draft variations are large, the height of the side opening must be increased accordingly. In these circumstances it is usually necessary for one of the more complex types of side door to be fitted. In long ships, large side doors are not normally situated in the more highly stressed region of the hull, roughly the centre third of its length.

Typical side ports weigh in the region of 0·25 tonnes/m² and loading platforms are similar.

6.7 Bulkhead doors

6.7.1 General

Doors are necessary to provide access through transverse bulkheads within a ship's hull. They are also needed at the weather deck ends of trunked ramps which are now found in many Ro-Ro vessels. When fitted below the freeboard deck, they must be watertight, but this is not essential elsewhere.

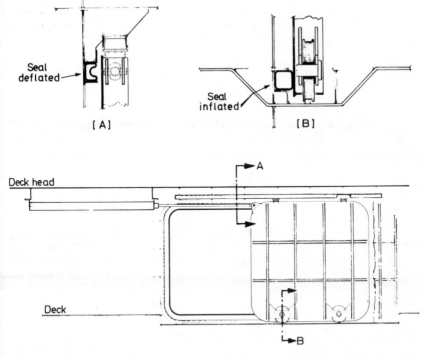

Fig. 6.50. Sliding bulkhead door, used to provide access through watertight bulkheads. Details A and B show the wheel arrangement and inflatable seal.

6.7.2 Types of bulkhead doors

6.7.2.1 *Sliding door*

Sliding bulkhead opening on rails as shown in Fig. 6.50 with their upper edges guided by rollers. Watertightness is assured by a rubber gasket against which the door is secured in the closed position. Inflatable gaskets are also used (Fig. 6.50) to secure the door, together with manual cleats. Sliding doors may be sill-less to facilitate the passage of wheeled vehicles. Whether hydraulically, pneumatically or electrically driven, they can be made almost any reasonable size, provided that sufficient stowage space is available. Sliding doors are used where it is impracticable to use side swinging doors.

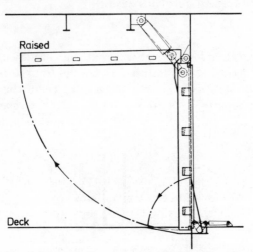

Fig. 6.51. Upward-pivoting door. This type of bulkhead door requires space to be clear of cargo for it to open and close.

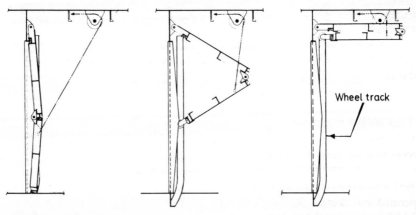

Fig. 6.52. Folding upward-pivoting door, which requires less clear space to operate than the type in Fig. 6.51.

6.7.2.2 *Upward pivoting door*

The upward pivoting door illustrated in Fig. 6.51 simply swings up to stow parallel to the deckhead. A small separate panel swings down to cover the recess in the deck to allow the smooth passage of vehicles. Rubber gaskets are employed to secure weathertightness. Hydraulic or manual cleats may be used and actuating mechanisms are also hydraulically operated.

A variant is the two-panel upward folding door which operates in a similar way but also folds at its mid-height. The bottom edge of the lower panel is guided by wheels that run in a vertical track (Fig. 6.52). This arrangement requires less clear space when being operated than the single panel door. As with some other doors, the threshold remains flush.

Fig. 6.53. Side-hinging door, used where large access openings with flush sills are required. The unhinged side of the door is supported by a wheel which rolls across the deck.

6.7.2.3 *Side-hinged door*

This type of door is used for large openings such as are required to give access to large vehicles in Ro-Ro vessels. It swings about a vertical hinge at one side as illustrated in Fig. 6.53, and its outer edge is supported by a single wheel which rolls across the deck as the door is opened. Usually hydraulically operated and cleated, this type of door is sill-less, providing a flush threshold, and is made watertight by means of a gasket that is pressed against a plate set into the deck.

6.7.2.4 *Weather deck door*

This is found in Ro-Ro vessels having an internal ramp leading to the weather deck. Weather deck doors fold upwards as shown in Fig. 6.54. They are situated at the fore end of superstructures or trunked ramp housings, and are hydraulically operated and cleated. They are usually provided with small ice-breaking jacks for use in severe weather conditions. They are sealed by rubber gaskets in the door which are pressed against the door frame during the closing and cleating process.

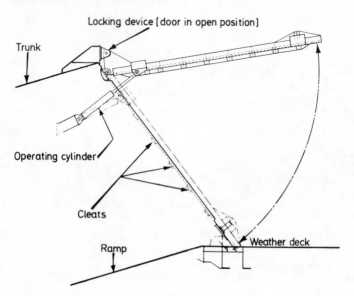

Fig. 6.54. Weather deck door, used to give access to the weather decks of Ro-Ro ships from an internal vehicle ramp. Containers can be stowed on top of the trunk.

6.7.3 Applications

Bulkhead doors are currently only fitted where they are essential, below the freeboard deck or on the weather deck. When fitted in the main vehicle spaces, they inevitably encroach on cargo stowage space and can impede cargo handling. They provide greater security against fire and flood, however.

If more stringent regulations governing Ro-Ro vessels are introduced in the future, as suggested in Sections 4.3.5 and 11.2.5 this could mean increased internal subdivision and more bulkhead doors. Bulkhead doors are designed to be equal in strength to their surrounding structure, and their cleating systems must also be of equivalent strength. They are usually hydraulically operated. Typical bulkhead doors weigh approximately 0·25 tonnes/m², depending on the water pressure head they are designed to withstand.

6.8 Car decks

6.8.1 General

Except in pure car carriers, car decks are usually designed to be movable so that they may be stowed away when not required to give more headroom for other vehicles. They are fitted in passenger/vehicle ferries and Ro-Ro vessels, where they are often used during the summer for tourist cars but raised in winter so that the vessels can concentrate on the carriage of commercial vehicles.

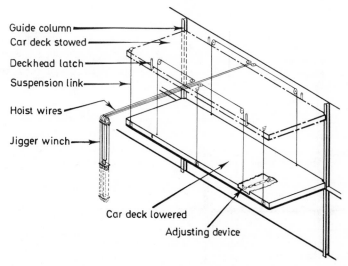

Fig. 6.55. Hoistable car deck, which can be hoisted by wires against the deckhead when not in use, to give increased headroom below.

6.8.2 Types of car deck

6.8.2.1 *Hoistable car decks*

The most commonly used portable car deck is the hoistable type. It can be raised against the deckhead by wires powered by a hydraulic jigger winch when not required, and is shown in Fig. 6.55. The hoistable car deck is locked in position by automatic latching devices. These decks are extremely light and may consist of a steel frame covered with plywood decking. The maximum axle load they can support is usually only ¾ to 1 tonne, for vehicles having an all-up weight of approximately 1½ tonnes (export cars) to 2 tonnes (tourist cars).

A form of multiple level hoistable car deck, illustrated in Fig. 6.56 is suitable for bulk carriers.

6.8.2.2 *Folding car decks*

A more complicated arrangement than the hoistable deck is the folding deck. This is illustrated in Fig. 6.57 and consists of three hinged panels which fold against the side of the hull and the deckhead by a directly-coupled hydraulic ram. The inboard end of the car deck is supported by a fixed wire, chain or bar as shown. This deck is constructed in a similar way to the hoistable type to keep its structure light, and to reduce stability problems in small ships. A simpler folding car deck is the pivoting type. This is usually one lane wide and stows vertically against the ship's side or casing, similar to the left-hand leaf in Fig. 6.58.

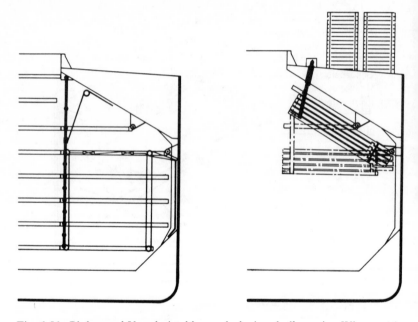

Fig. 6.56. Blohm and Voss hoistable car decks in a bulk carrier. When not in use the decks are hoisted up and swung out of the way as shown, allowing the hold to be used for cargoes like grain. The centre pontoons are lifted out by ship or shore cargo gear and stowed on the deck when not in use.

6.8.2.3 *Multi-folding car decks*

This arrangement divides a hold so that it can accommodate cars on two levels over its entire area (Fig. 6.58). The side sections hinge down as shown and the centre section is a standard multi-folding, hydraulic or wire-operated tween-deck hatch cover such as is described in Section 5.3. This arrangement is more often used in larger ships than the two systems described earlier. It is more robust and can support the greater number of cars carried in such vessels. But access to these decks is vertical, employing shipboard or shore cranes.

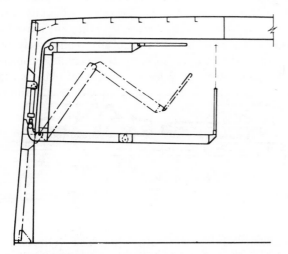

Fig. 6.57. Folding car deck which is folded out of the way when not in use, against the side of the vessel.

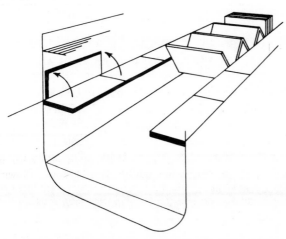

Fig. 6.58. Multi-folding car decks, used on larger ships where the car decks are required to be more substantial as they support a greater quantity of vehicles. The centre sections are of normal tween-deck hatch cover types, while the side sections hinge upwards.

6.8.2.4 *Sliding decks*

These are exactly the same as the sliding tween-deck hatch covers described and illustrated in Section 5.8. They are usually installed inside the hatch coamings of general cargo ships as shown in Fig. 6.59 and provide a useful space for cars in situations where they cannot be damaged by cargo being worked around them.

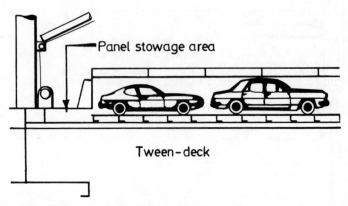

Fig. 6.59. Sliding car decks. This development allows cars to be stowed within the coamings of general cargo vessels, clear of other cargo and less prone to damage as they are loaded last and discharged first. Fig. 5.53 shows such a deck from above.

6.8.3 Applications

Hoistable and folding car decks are used in ferries and Ro-Ro vessels, while the multi-folding variety are usually installed in larger ships with big open holds. Bulk carriers are sometimes employed to carry cars (see Fig. 6.56) on return voyages which would otherwise have to be made in ballast, or when no cargo is available in the ship's usual trades. Such ships can carry up to 3000 cars, but they are being displaced on some routes by pure car carriers.

A typical ferry installation is shown in Fig. 6.60, where car decks for three lanes of vehicles are used on one side of the centre-line. They may be raised into their stowed positions on the other side to allow two lanes of commercial vehicles to be stowed beneath them.

Portable car decks are designed for vehicles weighing up to about 2 tonnes with a maximum axle loading of 1 tonne. This gives a deck loading of approximately 0·2 tonnes per square metre. Typical tourist cars require about 1·9 m clear height to allow passengers to get in and out and to accommodate roof-racks etc. Car deck structures are between 0·2 and 0·4 m deep, and weigh about 0·09–0·12 tonnes/m².

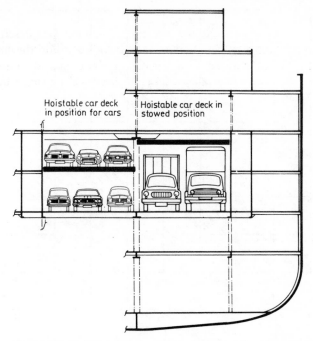

Fig. 6.60. A diagrammatic representation of the arrangement of a passenger/vehicle ferry's car decks in their lowered position with three lanes of cars and stowed to allow two lanes of commercial vehicles below.

6.9 Dimensions and weights

The dimensions of access equipment depend on the trades in which it is used together with vehicle weights and dimensions. Some typical figures are given in the preceding sections, and further detailed discussion is to be found in Chapter 8 which deals with specific ship types.

Access equipment weights vary with the loads supported, the openings spanned and their detailed structural design. Nevertheless approximate figures suitable for preliminary ship design purposes are available, since in practice axle loads do not vary widely for a particular type of ship. Some individual figures have been given in preceding sections and Fig. 6.61 gives some general guidance on ramp weights; more precise figures can be supplied by equipment manufacturers as required. The weights do not include the additional stiffening to be built into the ship to support the equipment. While this may amount, in quarter ramps, to as much as 30% of the ramp weight, for most other items of Ro-Ro equipment 10% is more typical.

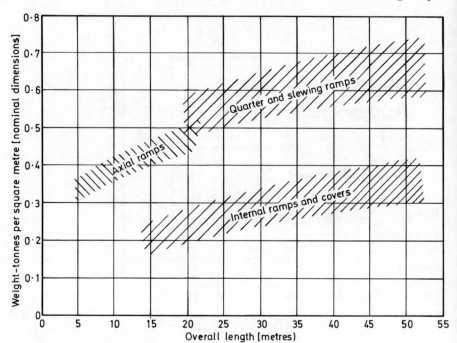

Fig. 6.61. Approximate weights per square metre of axial ramps, quarter ramps and internal ramps carrying typical Ro-Ro ship loads. The precise weight will vary with detailed design, width, supports and axle loadings. The nominal area is the overall length times the clear width of the narrowest section.

References

[6.1] Lane, P. H. R. The Design and Use of Shipboard Ramps. *Conference papers, Ro-Ro 77*, Business Meetings Ltd., 1977.
[6.2] Peckham, R. *Ship Access – The Key to Fast Turnround.* Cargo Systems International, December 1976.
[6.3] Hanson, P-A. Cost Implications of the Choice of Ro-Ro Equipment. *Conference papers, Ro-Ro 77*, Business Meetings Ltd., 1977.
[6.4] Taylor, G. R. Ro-Ro ships – State of the Art in Australia. *Marine Technology*, **13**, No. 4, October 1976.

CHAPTER SEVEN

Ship Design and Selection of Access Equipment – Bulk Cargo

Summary

This chapter discusses the considerations to be taken into account in the design of ships for bulk cargoes, their cargo spaces and the selection of their access equipment. The main types of bulk carrier are described, with the requirements for hatch sizes and types. A check list for choosing possible hatch covers summarizes the main considerations. A method of economic evaluation that includes more than first cost is described.

7.1 Access equipment for ships carrying dry bulk cargoes

7.1.1 Range of bulk cargoes

As described in Section 2.2, dry bulk cargoes are basically homogeneous commodities moving in complete shiploads, as opposed to general cargoes, which comprise a miscellany of smaller consignments. Bulk cargoes can be conveniently classified as follows.

  (i)   Major bulks – cargoes like iron ore, coal and grain, with worldwide shipments of over 50 million tonnes or more a year.
 (ii)   Minor bulks – cargoes like bauxite, sugar and salt, with annual shipments of between say 5 and 50 million tonnes.
(iii)   Semi-finished – cargoes like forest products and steel, which increasingly move in shiploads.

All bulk carriers are single-deck ships, with a double bottom, machinery and accommodation aft, and vertical cargo access through hatches. Speed is nearly always in the 13–16 knot range. A number of principal types are identified below.

7.1.2 Large bulk carriers

Large bulk carriers are ships of over about 50 000 tdw, now roughly 300 in number worldwide, engaged mostly in carrying large quantities of the major bulk commodities such as iron ore, coking coal and, to a lesser extent, grain, between specialized deepwater terminals. Such cargoes are normally loaded by gravity from overhead chutes and discharged by grab, so that large hatches are needed. A typical cross-section is shown in Fig. 7.1, with hoppered holds and sloped topside wing tanks providing a self-trimming*

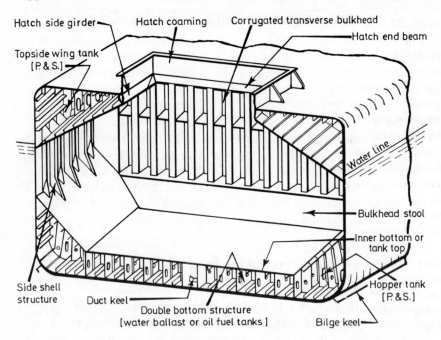

Fig. 7.1. This cross-section of a typical bulk carrier shows the classic arrangement of hoppered holds and sloped topside wing tanks which are well suited to the carriage of granular-type cargoes loaded by chute and discharged by grabs. The dimensions of the hatchway are suitable for either end-stowing or side-rolling types of cover.

capability as well as water ballast spaces. The number of holds varies between seven and eleven – usually an odd number so that dense cargoes like iron ore need only be loaded in alternate holds. This arrangement not only simplifies loading and discharge, but improves seakindliness by limiting metacentric height†; it does however require special attention to be paid to the influence of cargo distribution on longitudinal bending moments and shear forces. The total cargo hold length is about 78% of the ship length between perpendiculars. Sufficient cubic capacity is generally available to

---

* The shape is such that granular type cargoes can be loaded by gravity without having to be trimmed out into the wings of the hold.
† Metacentric height (GM) is a measure of a ship's stability when upright. The position of the centre of gravity (G) must be arranged to be below that of the metacentre (M) which depends on hull shape.

carry a full deadweight of coal or the heavier grains, stowing at about
$1\cdot2-1\cdot3$ m³/tonne (43–47 ft³ per ton).

The cargo length is usually divided into holds of equal length, except No. 1
(the foremost) which is often shorter for reasons of subdivision. Such ships
are typically designed for $B-60$ freeboard (see Section 4.1.2) so they must be
able to withstand flooding of any single cargo hold without submerging the
upper deck. In the popular 'Panamax' size ships, those with breadths up to
$32\cdot2$ m (106 ft), the limit of the Panama Canal locks, and ranging up to about
80 000 tdw, seven holds are generally required, each with maximum length of
about 25 m (82 ft). Sometimes two or three holds may be made longer than
the others, with intermediate holds shorter. Such an arrangement can give
greater flexibility for loading combinations of cargoes, or for reducing the
number of holds required to carry a full deadweight of iron ore. One or more
holds can be arranged for carrying water ballast, for improving seakeeping in
the ballast condition by deeper immersion, or for reducing air draft in port.
Such holds are usually only partly filled with water. As liquid 'sloshing'
pressures on bulkheads can be high, it may be necessary to adjust the length
of the hold or the depth of water in it to reduce them. $B-100$ freeboard can
sometimes be arranged, but the ship must be able to withstand flooding of
two adjacent compartments.

The depth of hold in large bulk carriers is such that transverse bulkheads
cannot often be arranged to extend between tank top and upper deck without
additional support. Usually this takes the form of a near-triangular stool at
the bottom, and sometimes at the deck as well, to reduce the span of the
main section, which is normally formed of corrugated plating, as shown in
Fig. 7.1. Double-skin bulkheads are sometimes used for additional strength
at boundaries to ballastable holds. The stools not only provide structural
support, but also assist hoppering the lower part of the hold and self-
trimming in the upper part, as well as providing water ballast space. They
do however limit the maximum length of hatch that can be fitted, although
not to a serious extent.

Subject to the internal constraint of the ship's structure, and the external
constraint of stowage area for hatch covers on deck, hatchway sizes on large
bulk carriers are generally as large as possible so that the holds are readily
accessible by grabs. Hatchway width is usually between 40 and 50% of the
ship's breadth, averaging 45%. Hatchway length is usually between 58 and
70% of the hold length, averaging 64%, but in any given case length is
dictated by the ship's structural arrangement and frame spacing. For
continuity of strength, hatch end beams are often arranged in line with web
frames in the holds and plate floors in the double bottom. Classification
Society rules limit the spacing of such frames and floors, while the spacing of
ordinary frames is usually between 700 and 1000 mm. The choice of hatchway length therefore usually alters by discrete amounts, depending on the

arrangement of the proposed hull structure. The dimensions of large bulk carriers are such that ship breadth is greater than hold length, with the result that hatchway widths are usually greater than hatchway lengths. Typical values are 14–22 m wide and 12–16 m long. The choice of suitable hatch covers is discussed in Section 7.2.

7.1.3 Medium and small bulk carriers

There are many trades where the quantities of cargo moving and the port facilities available do not permit the use of large bulk carriers. Such trades include bulk cargoes like bauxite, phosphate rock, non-ferrous ores including manganese, copper and nickel, salt, sugar, elemental sulphur and petroleum coke. Less homogeneous cargoes like steel products (coils, bars, pipes, plates etc.), pig iron, scrap metal and forest products can also move in shipload quantities. Although special ship designs for forest products have been developed (see Section 7.1.6), the other cargoes listed above, together with grain and coal, now employ about 3000 medium and small general-purpose bulk carriers between about 15 000 and 50 000 tdw. Most ships fall into either the St Lawrence Seaway category, i.e. ships of about 22 000 to 27 000 tdw with a breadth below approximately 23 m suitable for Seaway transits, or the slightly larger size range from 32 000–36 000 tdw with loaded drafts of about 11 m. Such ships can be accommodated at a very large number of ports throughout the world; the smaller ships are essentially 'go-anywhere' vessels, the larger 'go-most-places'.

The general arrangement of a small bulk carrier is shown in Fig. 7.2, and is basically similar to that of the large ships. Owing to their shorter length, fewer holds, usually from four to seven, are necessary in small vessels. The fewer the holds, the lower the first cost, but some flexibility of cargo distribution is lost. It is also less easy to satisfy regulatory requirements, for example with respect to stability when loaded with grain, or subdivision for B − 60 freeboard. Cargo handling equipment is fitted to the majority of smaller ships (unlike the larger vessels). This consists of either simple derricks and winches (about 50% of ships), or deck cranes (about 25%), depending on the anticipated operational requirements. While such gear is usually less efficient for cargoes like ore or grain than shore-based equipment, it is useful for the less homogeneous cargoes like steel or timber.

The weather deck layout of small bulk carriers is relatively complex compared to that of the larger ships. Grab discharging requires wide hatches to avoid damage to coamings and to reach all corners of the hold. Hatchway dimensions are therefore relatively larger. Hatchway width is usually between 45 and 55% of the ship breadth, averaging 50%. Hatchway length is less constrained by bulkhead stools, as their smaller hold depths mean that a single span of corrugated plating is usually sufficiently strong. However, cargo handling gear is usually fitted between the hatches and even if it is not,

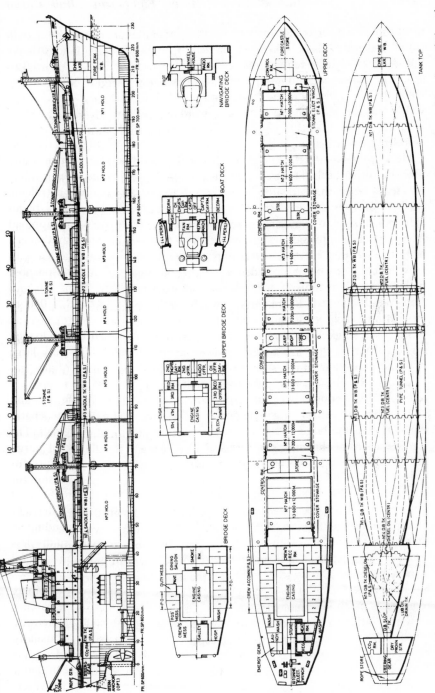

Fig. 7.2. This plan of Austin & Pickersgill's B26 26 000 tdw bulk carrier shows a typical arrangement of a vessel capable of transitting the St. Lawrence Seaway. Seven holds of varying length fitted with wire-operated single pull covers and served by derricks are illustrated, but alternative designs can have five holds, cranes or different types of hatch cover. No. 4 hold can be filled with water ballast. Nos. 2, 4 and 6 holds may be empty even if heavy cargoes are loaded in the other holds. Grain cubic capacity is 35 000 m³, speed 15 knots.

most ship designs make a provision for it. Deck cranes are often mounted on
cylindrical pedestals of 1·5–3·0 m diameter to improve operating coverage,
but derricks usually require masthouses to provide winch platforms and
space for switchgear and equipment. More elaborate cargo handling equipment such as gantry cranes travelling on rails impose limits on hatchway
width, and also influence deck shape forward in providing coverage for No. 1
hold. However, it is usually possible to arrange hatches of about 62–72% of
hold length, averaging 66%.

Deck space outside the line of hatches is required for ships likely to carry
timber or other deck cargoes; it may also be required for stowage of portable
car decks when they are not in use in the holds. The latter are fitted in
car/bulk carriers designed to carry up to 3000 cars on as many as eight
light, hoistable platforms suspended from the deck, as illustrated in Fig. 6.56.
Access hatches to topside wing tanks suitable for carriage of grain may also
be fitted. Where cargoes like steel rails or logs are anticipated, one or more
long hatches are often provided for easier stowage of awkward loads,
typically 16–20 m long. If such cargoes are expected to be more common
than grab-discharged bulk materials, twin hatches may also be more suitable,
the width of each being about 35% of the ship breadth.

Short-sea single deck cargo vessels may also be regarded as small bulk
carriers, because they are largely engaged in the transport of cargoes like
coal and grain. Paragraph ships predominate, i.e. ships built to a specified
value of GRT above which different regulations apply. 499 and 1599 GRT
vessels are very popular with corresponding deadweights of about 1300 and
3300 tonnes respectively in single deckers. One hold is usual in the smaller
types, two in the larger. Hatchways are generally as long and as wide as
considerations of ship strength, fore and aft access and hatch cover stowage
permit, to allow grab working. Hatch widths average 70% of ship breadth.

7.1.4 Ore carriers

The ore carrier is a specialized type of bulk carrier, designed to carry dense
ores, and although totalling some 300 ships at present, the type is much less
numerous than the ordinary bulk carrier. Iron ore is the commonest
commodity transported in them, so modern ore carriers are generally as large
as possible, over 100 000 tdw, to obtain the economies of scale. As iron ore
has a stowage factor of 0·4–0·5 m^3/tonne (14–18 ft^3/ton) the hold volume
required is much less than that for bulk carriers. Fig. 7.3 shows the cross-
section of a typical ore carrier, with its hold occupying only about 50% of its
breadth. The double bottom may be deeper than normal so that the centre
of gravity of the cargo is at a height giving a seakindly metacentric height.
The width of the hatchway need only be about 25–30%, of the ship breadth,
typically averaging 27%.

Because of the subdivision provided by the wing water ballast tanks, very long ore holds are common, usually only two, three or four even in ships of over 150 000 tdw. However each hold usually has two or three almost square hatches, so that all parts may be reached by loading chutes and discharge grabs.

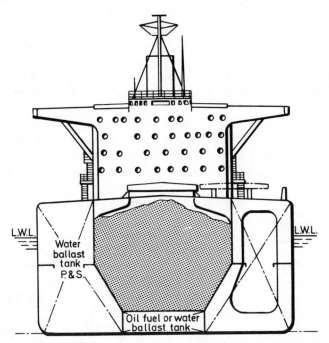

Fig. 7.3. The cross-section of a large ore carrier differs from a bulk carrier in that the main hold is much smaller, since the cargo is denser. The hoppered bottom assists grab discharge. A one-piece side-rolling cover is fitted, since hatch width is less than 30% of ship breadth. An ore/oil carrier would have a similar cross-section, except that the wing tanks would be used for cargo oil.

7.1.5 Combination carriers

The ore carrier configuration lends itself to a dual-purpose ship, capable of carrying ore in the centre holds on one voyage, and crude oil in the wing tanks on another voyage. Such ore/oil carriers have been used for over 50 years, but modern designs now also carry oil in the centre holds. Following the 1966 Load Line Convention, this extra capacity was required in order to load a full deadweight at the deeper draft permitted by $B-100$ freeboards. New designs of oiltight hatch covers and seals have made such developments possible. New anti-pollution regulations requiring minimum capacities of clean water ballast spaces may restrict the use of wing tanks for oil cargoes in the future.

The cross-section of an ore/oil carrier is the same as that of an ore carrier, but the centre holds are not quite so long. One or two additional transverse bulkheads are required, increasing the number of centre holds to four or five. Most holds generally have two hatchways, each having a length of about 33% of the hold length. Current modern ore/oil carrier sizes range typically from 120 000 to 300 000 tdw. The world combination carrier fleet consists of some 400 ships, including a number of smaller, older vessels.

In the same way, ore/bulk/oil (OBO) carriers are essentially large bulk carriers whose holds can carry either oil or dry bulk cargo. The general configuration and hatchway sizes are thus as in the large bulk carriers described in Section 7.1.2. Sizes range from about 70 000 to 180 000 tdw, rather smaller than ore/oil ships, since few cargoes other than iron ore are available in 150 000-tonne lots. While the dual-purpose capability adds about 10–15% to first cost, higher load factors can be achieved by 'triangular' trading (two loaded passages for every one in ballast), and advantage can be taken of higher freight rates by switching between the dry bulk carrier and the tanker market.

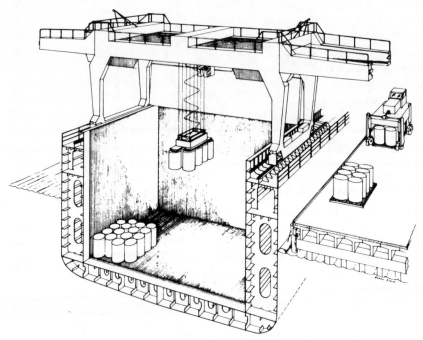

Fig. 7.4. This cut-away drawing shows a typical cross-section of an open or 'all-hatch' forest products carrier. The gantry crane is being used for loading paper reels. End-folding hatch covers are shown, but lift and roll covers could also have been fitted. Cargo can also be stowed on top of the covers. This type of ship, which can also be used to carry containers, was pioneered by R. N. Herbert with the 9300 tdw *Besseggen* in 1963.

7.1.6 Forest product carriers

Several specialized designs of bulk carrier have evolved for the transport of forest products – in total about 200 ships. The 'open' or 'all-hatch' ship is arranged so that when the hatch covers are opened, the entire hold is exposed. Thus a wide range of regular-shaped products like newsprint rolls, pulp bales or packaged lumber may be easily stowed. Fig. 7.4 shows a typical cross-section where the hatch width is between 75 and 85% of the ship breadth. In the fore-and-aft direction, only 1·2–1·5 metres separate adjacent hatchways. The hatches are very large and heavy, suitable for deck cargoes and the interiors of the holds are flushed off as square as possible. A double hull and double skin bulkheads provide a smooth interior as well as water ballast space. This arrangement also lends itself to the carriage of containers if cell guides are fitted. Usually one or two gantry cranes capable of discharging onto the quay span the hatches. Size ranges up to about 50 000 tdw, ships typically having five holds.

Another specialized type is the wood-chip carrier. Wood chips are a by-product of forestry operations which are used in pulp and paper-making, especially in Japan. As they have a very high stowage factor of 2·5–3 m³/tonne (90–110 ft³/ton) large hold volume is necessary. Thus these ships have a great depth and excess freeboard. Chip carriers range up to about 60 000 tdw, or 100 000 m³ capacity – equivalent to a normal bulk carrier of about 90 000 tdw. Internally the holds are hoppered at the bottom, but they do not have the bulk carrier's sloping topside wing tanks. Five or six holds are usual, with hatch widths about 50% of the ship breadth. Chip carriers are usually self-discharging, with grabbing cranes feeding hoppers on deck linked to conveyor belts running longitudinally outboard of the hatches onto a transverse conveyor forward for discharge ashore.

7.1.7 Other specialized bulk carriers

Other types of specialized bulk carrier include Great Lakes ships, cement carriers and continuous self-unloading ships. While special cargo handling equipment may be fitted, influencing the internal arrangement, e.g. conveyors at the bottom of the holds, the factors affecting cargo access equipment are generally covered in one or more of the ship types already described.

7.2 Choice of type of hatch cover

The selection of the appropriate type of hatch cover for vertical loading ships, including dry bulk carriers, is aided by a check list, such as the one below; detailed considerations under each heading are discussed in other sections.

7.2.1 Check list: Considerations in selecting hatch cover types

1. *Hatchway dimensions*

General considerations of ship type, number of holds (Sections 7.1 and 8.1 to 8.4)

(a) *Width: Minimum* from a consideration of cargo unit size, e.g. containers, and cargo handling equipment such as grabs.

Maximum from a consideration of the hull girder longitudinal and torsional strength, depth and weight of cover.

Multiple of any basic dimension, e.g. container width.

(b) *Length: Minimum* from a consideration of cargo unit size and cargo handling equipment.

Maximum from a consideration of hold length with respect to ship configuration; stowage space for covers; stowage requirements of particular cargoes, e.g. steel pipes, containers.

Multiple of frame spacing.

2. *Deck space for stowage* (Chapter 5)

(a) *End stowing*: Approximate clear hatch length to be available at forward and after ends of each hatchway.

(b) *Side stowing*: Approximate clear hatch width abreast hatchways.

(c) *Other types*: Lack of stowage area outside hatchways may justify consideration of lift and roll or pontoon types.

Personnel access past stowages.

3. *Clear height for stowage* (Chapter 5)

The presence of winch platforms or cargo handling equipment may limit clear height available for stowage.

Trade-off between length and height of stowage.

4. *Coaming height* (Section 4.5.1)

Minimum from statutory requirements.

Maximum from design and operational considerations.

Cargo volume of hatchways.

Adjustments to influence number and length of panels.

5. *Loading on cover* (Section 9.1.2)

Standard loadings from head of cargo and water (Table 9.1)

Distributed loads from timber or deck cargoes.

Point loads from containers or wheeled vehicles.

Flat top usually required for deck loads.

Stacking points for containers etc.

6. *Extent of opening* (Chapter 5)
Partial opening of hatch required or not.
Simultaneous opening of all hatches required (generally not possible with types like lift and roll).

7. *Operating mechanisms* (Chapter 5 and Section 9.6.2)
Deck cranes or winches available, and lead of wires.
Mechanical, hydraulic or electric actuation.
Fully automatic types.
Siting of controls.

8. *Weathertightness and cleating* (Section 4.1.5 and Chapters 5 and 10)
Importance of tightness; number of joints.
Manual or automatic arrangements.
Experience of claims for cargo damage etc.

9. *Operating requirements* (Chapters 3 and 4)
Opening and closing times.
Ease of operation.
Number and experience of crew.
Safety: visibility, fencing etc.

10. *Maintenance and repair* (M & R) (Chapter 10)
Maintainability of various types.
Number and experience of crew.
Space and time for access for M & R in port.
Cost of M & R: ship's crew, shore organizations.

11. *Cost* (Section 7.3)
Cost of bought-in equipment.
Cost of installation on board ship.
Influence on economics and performance of entire ship.
Trade-off of first cost versus operating advantages and costs.

12. *Weight* (Chapter 5)
Weight of cover as supplied.
Comparative weight as built into ship, including coamings, deck structure etc.
Effect on stability of ship, including height of deck cargoes stowed on cover.

13. *Construction and fittings* (Chapter 5 and Section 9.5)
Choice of material, e.g. higher tensile steel.

Single or double skin construction.
Fabricated size of components.
Delivery and installation arrangements with shipyard.
Special anti-corrosive coatings.
Choice of associated fittings required, e.g. ventilators.

7.2.2 Bulk carrier hatch covers

The types of cover suitable for a hatchway of given dimensions are broadly determined by the space available for stowage, with the final choice being influenced by the other considerations listed above, e.g. type of drive. Thus the majority of bulk carriers usually have the following types of cover.

7.2.2.1 *Large bulk carriers*
Predominantly side rolling pairs of peak-topped covers, because of their short wide hatchways (yet less than 50% of ship breadth). Usually they have hydraulic rack and pinion drive. Ships up to about 100 000 tdw may have single pull, sometimes of pan type.

7.2.2.2 *Medium and small bulk carriers*
Typically, these vessels have single pull covers, as hatchway length is usually greater than width. If stowage space is limited, folding covers are often used. Where cargo gear is fitted, direct pull covers are becoming more common. Roll stowing covers may be used where maximum weathertightness is required, e.g. for steel cargoes.

7.2.2.3 *Ore carriers*
These vessels are predominantly fitted with one piece side rolling covers.

7.2.2.4 *Combination carriers*
Usually combination carriers have side rolling covers, one piece in ore/oil carriers, two piece in OBOs (see Fig. 2.7).

7.2.2.5 *Forest product carriers*
Normally these have either lift and roll or flat-topped hydraulic end folding covers. Occasionally pontoon covers lifted by gantry crane are fitted. Lift and roll covers minimize the distance between hatchways, but they cannot all be opened simultaneously.

A short list of potentially suitable types can quickly be drawn up, but a detailed specification usually requires consideration of technical, economic and human factors. Hatch cover manufacturers are always ready to discuss these factors and put forward alternative design proposals. Fig. 7.5 shows sketches of the different ways a given hatchway can be covered, giving comparative information on weight and stowage space, assuming that there is no constraint on the latter.

7.3 Economic evaluation of alternative equipment

7.3.1 Comparing costs of access equipment

First cost is always an important consideration, but clearly other factors may indicate that the cover having the lowest first cost is not always the correct choice, especially if it requires significantly higher operating effort and time, or excessive stowage space encroaching on revenue-earning capacity, or maintenance and repair costs. Fig. 7.6 gives an approximate indication of relative costs for peak-topped wire-operated single pull covers. It can be seen that cost rises rapidly with span (but not in direct proportion), although there may be some compensation from the slightly reduced weight and cost of the hull structure outside the line of hatches. These costs only include the manufacturer's supply, and therefore exclude coamings, winches and installation on board. Installation costs can range between about 5% of hatch cover cost for simple types like pontoons, to about 20% for more complex types like side rolling rack and pinion. While the simplest type of single pull cover is cheapest, the addition of self-powered drive systems or automatic cleating soon bring the cost up to that of folding or roll stowing covers, as shown in Table 7.1. For very large spans, other types are often

---

Fig. 7.5 (*see next page*). The comparative stowage requirements and weights of a variety of hatch covers are shown, together with typical minimum coaming heights, with dimensions in millimetres. All clear openings are 13 m × 11 m, large enough for two 20 ft containers fore and aft, and four athwartships. Minimum regulatory loading of 1·75 tonnes/m² is assumed. The cover top plate dimensions are 13·25 × 11·30 m in all cases except side-rolling (*f*) (13·35 m), end-folding (*h*) (11·35 m) and roll-stowing (*e*) (13·46 m × 11·7m).

Figs. 7.5 (*a*) to (*d*) show single pull covers, flat-topped and peaked. Other arrangements of panel stowage are possible, e.g. 3 and 4, or 7, but are heavier than those shown. Vertical stowage dimensions for covers stowing aft are slightly lower than for covers stowing forward, as lower stowage rails are required to overcome trim aft of up to 2°.

Weights include the covers only, not coamings or stowage supports (see Section 5.2.8). Extra weight to support two-high containers on flat-topped single pull covers is about 25%, on pontoon covers (*k*) about 20% and on rolling or lift and roll (*j*) about 10%. Direct pull (*i*) and end-folding (*h*) covers are of such a design that their strength is sufficient to support two tiers of containers without extra reinforcement. The relative weights can vary with different hatchway dimensions, which may lead to more economical panel sizes for certain types of cover, e.g. end-folding.

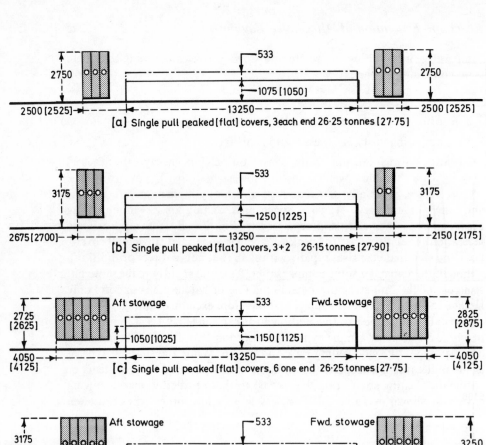

[a] Single pull peaked [flat] covers, 3 each end 26·25 tonnes [27·75]

533
1075 [1050]
2750
2750
2500 [2525]
13250
2500 [2525]

[b] Single pull peaked [flat] covers, 3+2   26·15 tonnes [27·90]

533
1250 [1225]
3175
3175
2675 [2700]
13250
2150 [2175]

Aft stowage   533   Fwd. stowage
2725 [2625]
1050 [1025]
1150 [1125]
2825 [2875]
4050 [4125]
13250
4050 [4125]

[c] Single pull peaked [flat] covers, 6 one end 26·25 tonnes [27·75]

Aft stowage   533   Fwd. stowage
3175 [3150]
1250 [1200]
1325 [1300]
3250 [3250]
3700 [3775]
13250
3700 [3775]

[d] Single pull peaked [flat] covers, 5 one end 26·15 tonnes

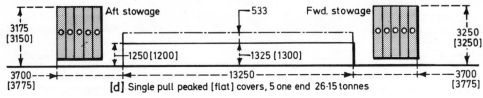

660
1330
2790
13460
2825

[e] Roll-stowing, 34·30 tonnes

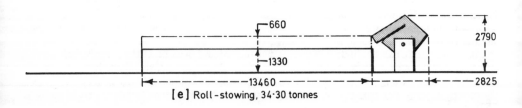

735
850
1685
5875
11300
5875

[f] Side rolling peaked, 24·0 tonnes

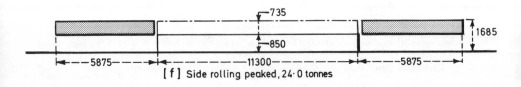

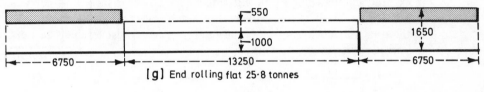

[g] End rolling flat 25·8 tonnes

6750 — 13250 — 6750

550, 1000, 1650

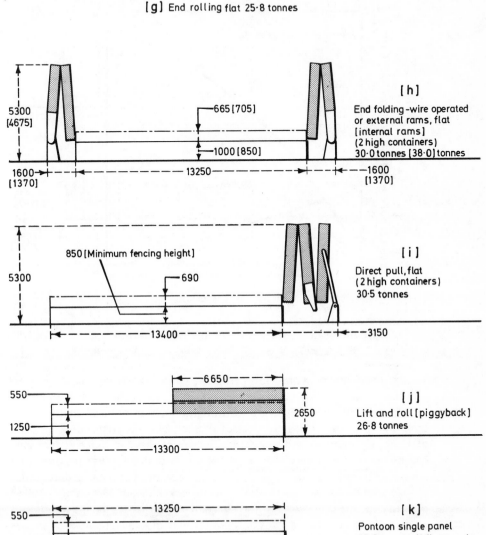

[h]
End folding – wire operated or external rams, flat [internal rams] (2 high containers) 30·0 tonnes [38·0 tonnes]

5300 [4675], 665 [705], 1000 [850], 1600 [1370], 13250, 1600 [1370]

[i]
Direct pull, flat (2 high containers) 30·5 tonnes

5300, 850 [Minimum fencing height], 690, 13400, 3150

[j]
Lift and roll [piggyback] 26·8 tonnes

6650, 2650, 550, 1250, 13300

[k]
Pontoon single panel 25·0 tonnes, if five panels 27·75 tonnes

13250, 550, 850

Panel stowed on quay

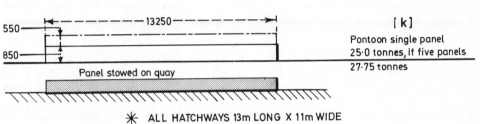

✳ ALL HATCHWAYS 13m LONG X 11m WIDE

more practical than single pull, e.g. side rolling covers in large bulk carriers.

For ordinary bulk carriers, the cost of hatch covers is usually about 2–3% of the total ship cost, but for specialized types, like forest products carriers, the proportion can rise to 5 or 6%.

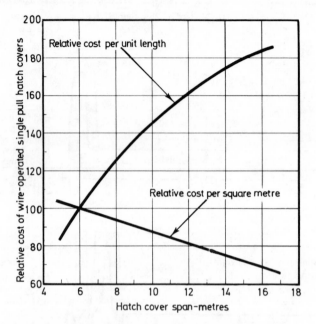

Fig. 7.6. This graph shows the relative cost of wire-operated single pull hatch covers. Although cost increases with span, the unit cost per square metre falls, as the mechanical parts do not increase much in proportion to size.

Like most other engineering products, cargo access equipment can be purchased either in a basic functional form, or with a more elaborate specification to yield operating advantages such as reduced maintenance cost. The problem then is to decide whether it is worth spending more in first cost in order to save more over the life of the ship. Such judgements can be made in a variety of ways, from the instant subjective decision to a complete techno-economic study. Some form of economic evaluation is always useful, not necessarily to indicate automatically the choice, but to quantify the life-cycle costs of the alternatives so that the final decision can be made based both on a knowledge of estimated financial figures as well as on less quantifiable grounds like safety or operator convenience.

Where economic evaluations are concerned largely with trading off increased first cost with reduced operating costs, without any impact on the ship's freight-earning performance, a very simple form of calculation is usually sufficient. The following example outlines this approach; for more

Table 7.1. Relative costs of different types of hatch cover

Type of weather deck cover	Approximate relative cost
Wire-operated single pull, peak-topped	100
ditto, flat-topped	105
ditto, for two-high container loading	130–140
Single pull, fixed chain drive	130–135
ditto, plus automatic cross-joint	140–145
Single pull, fixed chain drive, automatic cross-joint, hydraulic lifting and locking	180–190
Direct pull	115–120
Hydraulic folding, external mechanism	140–160
Side rolling	150–160
Roll stowing	170–190
Pontoon cover, one piece	90–100

Type of tween-deck cover	Relative cost (*basis about* 70% *of weather deck*)
Pontoon cover	100
Sliding sections, wire-operated	105–110
Multi-folding wire-operated	115–125
Hydraulic folding	170–180

detailed discussion of the principles of economic calculations, see elsewhere [7.1].

7.3.2 Example of simple economic evaluation

Uncoated mild steel compression bars on cross-joints of single pull hatch covers are liable to corrode and require replacement several times during the life of the ship. Leakage may result before repairs can be undertaken. A compression bar may be made of stainless steel round bar; it will then normally last the life of the ship.

The extra cost of such a bar, fitted on all cross-joints of a Panamax bulk carrier's hatches is estimated as £5000 on a basic price of £200 000. It is considered that renewals of all mild steel cross-joint compression bars will be required after 8, and 16 years' service, at a present cost of £4500 each time, and it is assumed that replacement costs rise by 10% each year.

Does the more expensive compression bar look a good investment, for example yielding a rate of return of say at least 15%?

The basic approach to such calculations is to recognize that money spent or received in future years has lower 'present worth' than that same sum now. A sum of say £4500 in 8 years' time could have been accumulated by depositing a smaller sum, say £2000, in a bank now, which under compound interest would have increased by a further £2500 in 8 years. The actual interest rate, i, (as a decimal fraction) is given by the following relationship: Future sum of money = Present sum of money $\times (1 + i)^N$ where N is the number of years.

In this case $(1 + i)^N = 4500/2000 = 2{\cdot}25$

For $N = 8$, $i = 0{\cdot}107$ or $10{\cdot}7\%$ (from tables [7.1] or calculating $(2{\cdot}25)^{1/8} - 1$, using y^x function on a pocket calculator).

Any future sum of money can thus be expressed in 'present worth' terms by multiplying it by the 'present worth factor' $1/(1+i)^N$.

The annual cash flows in the example can be sketched in the pattern illustrated in Fig. 7.7. The problem is to calculate whether the present worth of the replacements over the ship life is greater than the initial extra investment, when discounted at 15%.

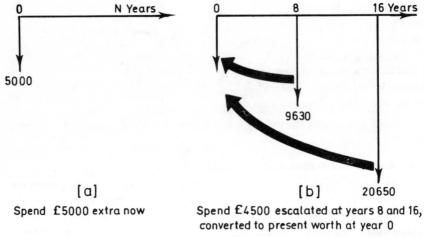

[a]

Spend £5000 extra now

[b]   20650

Spend £4500 escalated at years 8 and 16, converted to present worth at year 0

Fig. 7.7. This sketch shows the timing of cash flows when comparing an increased investment now with higher replacement costs in future years (as described in Section 7.3.2). Table 7.2 shows how the cash flows are converted to present worth terms.

A table such as Table 7.2 can be drawn up to calculate the present worth of the cash flows in Fig. 7.7(b), by multiplying each year's cost by the present worth factor, i.e. column 4 = column 2 $\times$ column 3. Column 2 shows the replacement cost escalated at 10% per annum, i.e. increased by $(1{\cdot}10)^8 = 2{\cdot}14$ times by Year 8.

Table 7.2. Present worth cash flow calculation

Year (1)	Replacement cost £ (2)	Present worth factor (3)	Present worth (4) £
8	$2 \cdot 14 \times 4500 = 9630$	$(1 \cdot 15)^{-8} = 0 \cdot 327$	3150
16	$4 \cdot 59 \times 4500 = 20650$	$(1 \cdot 15)^{-16} = 0 \cdot 107$	2210
	30280		5360

Thus it can be seen that even after discounting the large sums in Column 2 to Year 0 (time now), the present worth of the mild steel bar (£5360) is higher than that of the more expensive one (£5000), so that the latter is the better investment, giving in effect a capital gain of £360, even before taking any credit for the lower likelihood of leakage and reduced time out of service. Alternatively the calculation could have been carried out in real rather than in money terms by ignoring inflation and using a lower required rate of return.

Where a choice of equipment can influence the ship's annual earning capacity, e.g. by reducing port time, the evaluation becomes more complex. In addition to any difference in first cost of the hatch covers and possibly financial terms associated with them, there may well be differences in annual income and expenditure deriving from differences in the following;

(i) cargo payload, due to weight and dimensions of alternative covers.
(ii) port time, due to ease of cargo handling or operating times for covers.
(iii) crew requirements for operating covers, reducing overtime or permitting the ship to be more readily operated with minimum crew.
(iv) maintenance and repair costs, due to better materials, specification and maintainability.
(v) weathertightness, and associated cargo claims.

Estimating the financial differences due to such factors is not easy, being subject in many cases to the operator's judgement. As discussed in Section 3.3, savings in time due to improved equipment can be translated into money terms by use of long term time charter freight rates. Differences in payload may also be handled in this way, so that a ship which is a deadweight carrier (e.g. a bulk carrier), whose equipment is say 30 tonnes lighter than the alternative, may assume additional annual earnings of:

$30 \times$ time charter rate per tonne d.w. per month (say $5·00) $\times$ assumed months on hire per annum (say $11·5$) = $17 200

Examples of more detailed evaluations of alternative ship designs are given elsewhere [7.1].

Reference

[7.1] Buxton, I. L. *Engineering Economics and Ship Design.* British Ship Research Association, 1976.

CHAPTER EIGHT

Ship Design and Selection of Access Equipment – General Cargo

Summary

Design features of the principal types of vessel for carrying general cargo are described, along with the factors which influence the selection of their cargo access equipment. Section 8.1 discusses multi-deck ships for break-bulk cargoes and Section 8.2 container ships. The wide variety of Ro-Ro types are divided thus: Section 8.3.1 Passenger/vehicle ferries, Section 8.3.2 Short sea freight Ro-Ros, Section 8.3.3 Deep sea Ro-Ros. Section 8.4 discusses miscellaneous types including refrigerated vessels, barge carriers and heavy-lift ships. Finally, Section 8.5 discusses the choice of ship type for carrying general cargo. It is suggested that container ships are the primary deep sea unit load carriers, with Ro-Ro vessels providing flexibility for certain trades. Break-bulk vessels and barge carriers find their best applications serving low labour cost or developing countries.

8.1 Multi-deck break-bulk ships

As described in Sections 2.4 and 2.5, general cargo includes a wide range of manufactured, semi-finished and food products, either in break-bulk or unitized form. There is an equally wide variety of ships for carrying such cargoes, from the traditional tween-decker to the highly advanced Ro-Ro.

8.1.1 Conventional ships

The long-established, conventional break-bulk cargo vessel (or freighter) with two or more decks continues to be built in large numbers, especially for trading to developing areas of the world like South America or Africa. The present world fleet of multi-deckers of more than 1000 GRT is about 10 000 ships, although only about 3000 of them can be classed as 'cargo liners'. The principal deep sea types currently being built range from the Liberty replacement type, carrying a deadweight of about 15 000 tonnes (bale capacity 19 000 m³ (670 000 ft³)) at a speed of about 15 knots, to the complex,

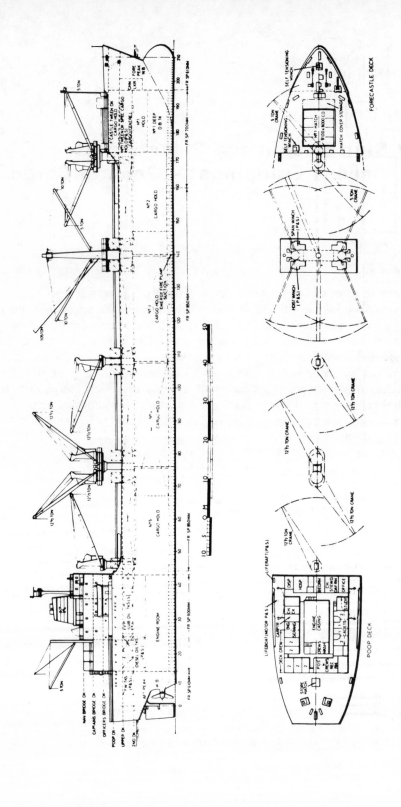

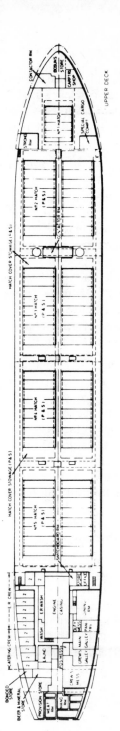

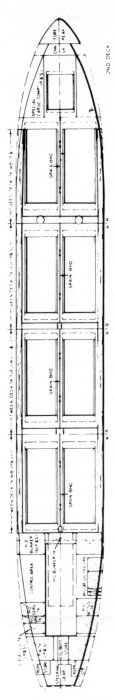

Fig. 8.1. This general arrangement of a modern cargo liner shows one of the Govan-built 23 000 tdw 16-knot vessels for Kuwait. Although rather larger than most ships with 29 000 m³ bale capacity, the general features are typical. Five holds, each with one tween-deck, are accessed by twin hatches, except No. 1, where the ship is too narrow. Each hatchway can accommodate three 20 ft containers fore and aft and three abreast, three-high in the lower hold, two-high in the tween-deck. Each hold is divided by a corrugated centre-line bulkhead. Weather deck hatches consist of chain-driven single pull covers, stowing at both ends, while tween-deck covers are wire-operated folding covers, whose stowage space is indicated on the plan. The 105-ton derrick is suitable for handling cargoes such as heavy machinery, military vehicles like tanks or deck-stowed cargoes like lighters. Fig. 2.1 shows one of these vessels loading cargo.

semi-container tween-deckers of 25 000 tdw, 30 000 m³ bale (1 060 000 ft³) and 20 knots. Smaller vessels, up to about 6000 tdw (8000 m³ bale), are used in more localized trading areas like the Mediterranean or the Far East. The design characteristics of the most popular larger types such as that illustrated in Fig. 8.1, can be summarized as follows:

(i) four or five holds, to provide reasonable cargo access and distribution for multi-port itineraries;

(ii) bridge and machinery aft, or with only one hold abaft them. The latter arrangement is more appropriate to the faster, fine-lined vessels which can encounter trim and machinery space problems if 'all-aft';

(iii) one tween-deck, usually deep enough to stow containers two-high in the hatchway;

(iv) full outfit of cargo gear, ranging from multiple pairs of 5-tonne derricks at each hold in the simpler ships, to 20–25-tonne cranes capable of lifting containers in the more advanced ships;

(v) overall dimensions selected to suit a large number of general cargo berths worldwide, with normal draft less than 9·5 m (31 ft).

The requirements for cargo access are governed by the need to load a large number of small consignments, often consisting of irregularly shaped packages. Thus weather deck hatchways are made as large as can conveniently be arranged, so that cargo can be accurately 'spotted' over a large area of the hold. This reduces the amount of handling out to the wings and overstowage (cargo for later ports of call stowed on top of cargo for earlier ports). Twin hatches abreast in the midship region are often adopted. Tween-deck hatchways are usually the same size as those on the weather deck, and are always flush to permit the operation of fork lift trucks – a major change from a generation ago.

With the widespread adoption of fully unitized services on many deep sea routes, the rather elaborate designs of the mid and late 1960s are rarely built now. In particular, triple hatches abreast are no longer fitted; apart from the cost and the complicated deck layout and structure, it has been found that interference between gangs and fork lift trucks prevents full advantage being taken of this arrangement. Two or more tween-decks are less often fitted than hitherto, as consignments on conventional cargo liner routes tend nowadays to include a large proportion of 'bottom' cargo, like steel or machinery, which needs more stowage space per unit. Delicate cargoes like television sets or toys require a lot of 'shelf space' and tend to move mostly on trade routes which are now largely containerized. Few general cargo ships now have deep tanks for carrying valuable liquid cargoes like vegetable oils or latex. These are now more often carried in specialized 'parcels' tankers. Part of one lower hold near amidships is sometimes arranged as a deep tank, suitable for either general cargo or up to about 2000 tonnes of water ballast to improve seakindliness when the ship is lightly loaded. Extensive re-frigerated space is rarely found in modern general cargo vessels, as 'reefer' cargoes now move almost exclusively in specialized multi-deck ships or in container ships (see Section 8.4.1).

With the increasing numbers of containers being transported by sea, even on routes which are not fully unitized, most modern cargo liners are designed around the container module. Even if not immediately required for a particular trade, this provides flexibility and second-hand saleability. Thus the typical modern cargo liner has five holds, one or two 27–30 m long suitable for cargoes like steel pipes. No.1 hold is usually shorter than the others, since its awkward shape slows down cargo handling compared with the midship holds.

The lower holds are also suitable for stacking containers three-high, so that depth from keel to tween-deck is usually about 9·5 m. The tween-deck height, from beam to beam is about 3·6 m. Thus, allowing for hatch coamings about 1·5 m high and deck camber, two 8 ft 6 in high (2·59 m) containers can be stowed within the hatchway, as shown in Fig. 8.2. Depth to the weather deck is at least 13 m (compared with a maximum draft of about 9·6 m). The hatches may be in pairs where the shape of the deck permits, each hatchway being about 35% of the ship's breadth and spaced 1–1·5 m apart transversely. Their actual dimensions must be multiples of about 2·6 m, adequate width for an 8 ft wide (2·44 m) container with appropriate clearances.

The breadth of a general cargo ship is governed largely by stability require-

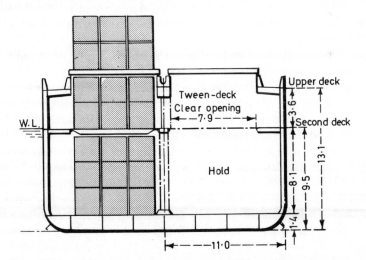

Fig. 8.2. A cross-section of a typical two-deck cargo vessel, whose dimensions are selected to allow containers to be stacked three-high in the hold and two-high in the tween-deck. Containers can also be stacked on the hatch covers, one, two or three-high. Typical dimensions are shown in metres.

ments; with a depth of about 13 m, it is likely to be at least 21–22 m. This lends itself readily to twin hatchways each about 7·8 m wide, and permits three rows of containers to be stacked in each. A partial or complete centre-line bulkhead is sometimes fitted in the holds, although pillars are normally adequate.

The length of the hatchways is usually a multiple of the normal 20 ft container length (actually 6·06 m) or the 40 ft length (12·19 m), although a minimum of about 6·5 m is necessary to allow for clearances and hatch corner curvature. Thus hatchways are typically about 13 m, 19·5 m or 26 m long, corresponding to two or three 20–40 ft container stacks. The actual length depends on the frame spacing. Hatchway length is usually between 55% and 75% of the hold length, averaging 63%, so that holds tend to be either from 18–23 m or 26–32 m long. The weather deck covers are usually arranged for stacking containers two-high, less commonly one or three-high. The total load may be restricted for three tiers with the result that not all the containers can be filled to their maximum allowable weight. Hatch covers must be flat-topped for containers; sometimes additional containers can be stowed outboard of the hatches.

If appropriate cargo handling gear is fitted, the modern cargo liner can be either a semi-container vessel capable of carrying 600 or 700 TEUs of containers, or a general cargo vessel with space for about 24 000 m³ of break-bulk goods. However, cargo deadweight does not usually exceed 12 000 tonnes, unless particularly dense semi-bulk cargoes are being carried on a particular voyage. If the ship has been assigned dual tonnages, the Tonnage Mark (Section 4.2.2) will not often be submerged, so allowing the lower GRT and NRT figures to be claimed.

Occasionally, single hatches may be fitted, each capable of stowing five or six containers abreast, with a width about 50–70% of the ship's breadth. Single hatches are more commonly found in the simpler Liberty replacement types, such as the SD 14, where the container is a less dominant factor in the design. For such ships, hatchway widths are more commonly 40–50% of the breadth, and hatchway lengths 50–60% of hold length.

The selection of an appropriate type of hatch cover must take into account the dimensional considerations mentioned here, as well as the factors listed in Section 7.2.1. Since deck stowage space is limited and the entire hatch must be openable, rolling, lift and roll, and pontoon covers are rarely fitted at the weather deck. Single pull covers are by far the most popular choice, stowing under overhanging winch or crane platforms sited between the hatches (see Fig. 8.1).

Wire-operated covers are the cheapest and are therefore normally included in the shipyard specification of standard designs. More advanced types, such as fixed chain drive or end folding covers, are becoming increasingly popular for semi-container ships, especially where rapid operation with small crews is required. No.1 hatch in particular is often fitted with end folding covers owing to the restricted dimensions and stowage space. Folding covers have also been used for midship holds. Direct pull and roll stowing covers are also gaining acceptance, particularly on wide single hatchways.

Tween-deck covers are operated less often than weather deck covers, as only the latter are usually closed after work in port has ceased for the day, or in wet weather. The most popular types are end or side folding, wire-operated in the simpler ships, electric or hydraulic in the more advanced. Other types in common use include pontoon covers (simple ships) and sliding covers. All these allow partial opening of the hatch and are usually designed for fork lift truck loading. Tween-decks and their covers are usually designed to take truck loadings of at least 5 tonnes on a 2-wheel front axle or 8 tonnes on a 4-wheel axle, corresponding to payloads of about 2½ and 4 tonnes respectively. Common pallets do not exceed these loads, but if heavier cargoes like steel products are carried in the tween-deck (as opposed to the hold with its much stronger double bottom) strengthening for trucks up to about 20 tonnes weight is necessary. Tween-deck covers are usually designed to take the loads imposed by 20 ft containers stacked two-high. This normally allows uniformly distributed loads of about 2·5–3·5 tonnes/m² to be carried. Loading on weather deck covers is usually less severe, as fork lift trucks do not operate on them; design loads range from the usual minimum 1·75 tonnes/m² up to whatever container weights are anticipated, typically stacks of 20 ft containers, each with a total load of about 30–40 tonnes (see Section 9.2.2).

8.1.2 Variants of the conventional multi-decker

Similar considerations to those discussed above also apply to small tween-deck ships, but on a smaller scale. Hatchways occupy a greater proportion of the ship's breadth, sometimes up to 80%, to give a reasonable clear opening 10–11 m wide. In some vessels, such as paragraph ships, the second deck is installed largely for tonnage purposes, with the result that tween-deck covers are not usually designed to carry significant loads like fork lift trucks.

Variants of the larger type of cargo vessel are sometimes built. While many cargo liners carry one heavy derrick of perhaps 100 tonnes capacity, there are a number of other heavy-lift ships designed with additional capabilities. For these a lifting capacity of 300–500 tonnes, serving a hold 45 m or more in length for awkward cargoes like large pressure vessels, is not unusual. Hatchway dimensions may thus be as large as 35 m by 10 m, with pontoon covers, handled by the ship's gear, capable of supporting heavy loads. Small specialized heavy lift vessels are discussed in Section 8.4.3.

Pallet-carrying multi-deck ships became popular in the 1960s. These were often fairly small vessels operating on coastal routes like those around Scandinavia. A few ships, often specialized vessels for forest products or refrigerated cargoes, were built for deep sea routes where cargoes are reasonably uniform. Such ships were fitted either with no hatches at all, or small hatches intended only for emergency use. The primary means for cargo access was side doors, with fork lift trucks on the quay landing their cargo on to a loading platform on the ship, whence it could be stowed either by other fork lift trucks inside the ship, or by elevators and conveyors (see Fig. 2.5 and 6.44). As described in Section 6.6.2, doors can be simple side ports or combined with deck edge hatches, 2–6 m wide and as much as 10 m high.

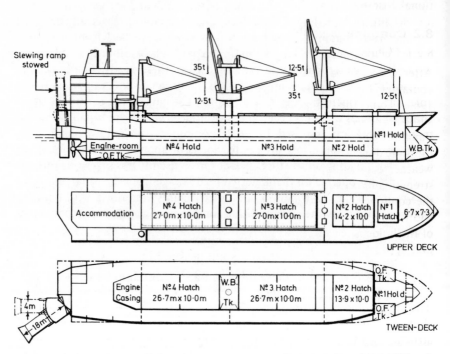

Fig. 8.3. The Hamlet Multi-flex design is a versatile 15-knot cargo ship capable of carrying break-bulk cargoes, bulk cargo, containers or Ro-Ro cargo, 13 000 tdw, 20 000 m³ bale. The tween-deck can be accessed by a slewing ramp. Weather deck and tween-deck hatch covers are of the folding type. The after section of No.4 hatch may be formed by a pontoon which can carry water ballast, and which can also be lifted into the water by the cranes to serve as a lighter. Where an operator requires a Ro-Ro capability and also the ability to handle the general range of break-bulk cargoes, the versatility of this type of vessel may be able to justify its additional cost.

Despite the apparently attractive economics offered by cargo handling rates as high as 200 tonnes per hour per side door, large pallet carriers have not gained wide acceptance on deep sea general cargo routes, partly through strong competition from other forms of unitized transport, partly through organizational problems which arise when thousands of pallets have to be marshalled in the less well-equipped ports of the world.

Pallets continue to be widely used in conventional cargo liners, and they are also suitable for the various newly developed, hybrid types of ships. Multi-deck designs with stern and/or side doors into the tween-deck, some-times with lower holds used for bulk cargoes, have been marketed and a few have been built. Fig. 8.3 shows one current design. Section 8.5 discusses the circumstances under which such additional flexibility may justify its addi-tional cost.

8.2 Container ships

8.2.1 Cellular ships

After a decade of pioneering by a few American operators, the cellular container ship was accepted on several major general cargo routes in the late 1960s. Today there are some 600 purpose-built container ships, ranging from short-sea vessels of about 100 TEUs capacity (about 1500 tdw) up to the Panamax giants of 3000 TEUs (50 000 tdw). The design philosophy is simple: the hull dimensions are geared completely to container modules and the weather deck is opened up to the maximum extent compatible with ship strength and deck layout requirements, in order to provide vertical access to the container cells. All containers are 8 ft wide (2·44 m) so that the hold width is a multiple of about 2·6 m to allow for the cell guides. The total width of the holds is usually between 79 and 85% of the ship's breadth. Thus as a first approximation, breadth in metres is

$$(3·1 \text{ to } 3·4) \times (\text{number of containers stacked abreast}).$$

The strips of deck abreast the hatches occupy only about 8–10% of the ship's breadth each side. As discussed in Section 9.1.4, careful design of this structure is necessary to obtain adequate longitudinal and torsional strength. No container ships have yet been built where breadth exceeds the maximum for the Panama Canal of 32·2 m (106 ft), so that the greatest number of containers stowed abreast below deck is ten. Four abreast is the smallest number of rows generally found in purpose-built container ships.

The depth of the hull is also a function of the container dimensions. Early ships were built around 8 ft high containers (2·44 m) stacked up to six high. According to ISO recommendations, containers must be strong enough to withstand being stacked six-high fully loaded, when the total static load could be 122 tonnes (120 tons) (i.e. six 20 ft containers at 20 tons each). But while some containers are loaded to their weight limit (and sometimes over), average container weight in practice is only about 12–15 tonnes. For 40 ft containers the average weight is between 15–20 tonnes, compared with an allowable maximum of 30·5 tonnes (30 tons). It is often possible therefore to stack eight or nine-high without exceeding the permissible load on the bottom container.

Early containers were only 8 ft high, but taller containers have since become more common in an effort to increase cubic capacity. 8 ft 6 in high (2·59 m) containers are the norm today for design purposes, although 9 ft (2·74 m) and 9 ft 6 in (2·89 m) have also been built. Since the last two are too high for the railway and highway loading gauges in many countries, they are used mainly on routes serving North America. Modern container ships are designed around the 8 ft 6 in high module, which permits a mix of 8 ft and 9 ft 6 in containers to be stowed as well. Allowing for the depth of the double bottom and some projection into the coaming space, the depth of the ship from keel to deck is usually a multiple of about 2·7 m.

The number of rows of containers abreast and tiers deep depends on the stability characteristics of the ship, taking into account the effect of containers stowed on deck. Common combinations for the larger ships are:

Rows wide		Tiers deep
10	×	9
9	×	7
8	×	6

Stowing box-shaped containers in a ship's hull inevitably means losing cargo space in the curved regions and outside the hatchways. Much of this lost volume can be regained by stowing on deck, as the containers are weatherproof. Container ships like the one shown in Fig. 2.14 are therefore designed to carry containers supported on the hatch covers, usually two, three or four tiers high depending on the stability of the ship, and the line of sight forward from the bridge. If the size of the ship permits, from one to three extra rows of containers can also be stowed abreast compared with below decks, since containers can be stowed on supports right out to the ship's side, and no space is needed for cell guides or coaming structure. A proportion of lightly-loaded or empty containers are usually stowed on deck, partly for stability reasons, but also to reduce the loading on the hatch covers. The net result is that 30–40% of the total number of containers carried are stowed on deck, as shown in Fig. 8.4.

Since every container is uniquely identifiable, together with its load, it is possible to plan the stowage ashore before the ship arrives in port, so that all the following requirements can be met;

    (i) overall centre of gravity to give adequate ship stability;
    (ii) total load in a stack within permissible limits;

  (iii) no overstowage of containers for different ports;
  (iv) ship trim and strength satisfactory;
  (v) separation of special containers, e.g. those filled with refrigerated or hazardous cargo.

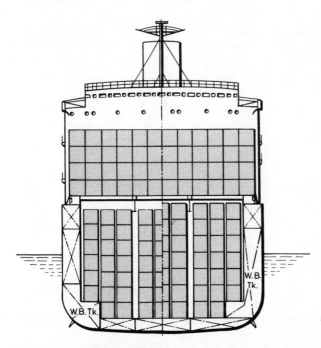

Fig. 8.4. A typical container ship cross-section, where the hatches cover 80% of the breadth. The left-hand section of the hull shows 8 ft containers stacked seven-high, the right-hand 8 ft 6 in containers. The three pontoon covers abreast can take containers stacked four-high, although not all at maximum weight. Oil fuel and water ballast are carried in wing and double bottom tanks.

Pre-planned stowing puts the heavy containers at the bottom of a stack and, compared with random stowing, increases the number of containers that the ship can carry in the vertical direction.

  Most container ships have their machinery right aft, except for some large multiple screw vessels where machinery space requirements dictate a ¾-aft position. They generally have between three and eight holds bounded by watertight bulkheads, but within each hold there may be one, two or three

Fig. 8.5. This photograph shows the container ship terminal at Tilbury, with the 1300 TEU *Act 5* at one of the berths. Three container cranes are alongside the ship, with the middle one working. Containers are brought to and from the crane by trailers, or by the straddle carriers seen in the left foreground. The majority of containers on the ship and in the park are 20 footers for the Australian trade. In the background can be seen a timber berth and beyond, the rail sidings where containers arrive by train.

stacks of containers, each stack consisting typically of two 20 ft or one 40 ft container, and all supported by vertical cell guides composed of 150 mm × 150 mm steel angles. Additional transverse stiffening is usually fitted between stacks, especially at the deck. The fore-and-aft distance between adjacent stacks may be as little as 0·9 m (3 ft) but it is more generally about 1·5–2·5 m (5–8 ft). Refrigerated holds have slightly larger overall dimensions to allow both for the insulation of the structure (to avoid low-temperature brittle fracture) and for the ducting systems required to circulate cooled air to each insulated container. Reefer containers which have individual integral refrigerating units can be carried on deck in ships without refrigerated holds. They are supplied with electric power by cable to run their machinery.

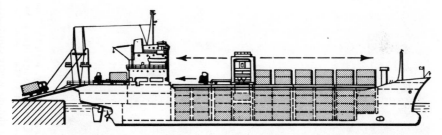

Fig. 8.6. A small self-sustaining 17-knot container ship of the *Strider* class is shown, of 6500 tdw or 330 TEU capacity. The angled quarter ramp allows trailers to be driven aboard at ports without full container facilities, across the hatch covers to where the 38-ton gantry crane can take their containers and stow them in the cellular holds. Two hatch covers abreast are fitted, each covering three containers wide and two long. They are of the pontoon type, handled by the gantry crane.

Containers are usually handled at specialized terminals by shore-based cranes lifting from ship to trailer or straddle carrier for delivery to their individual storage locations as illustrated in Fig. 8.5. Shipboard gantry cranes are occasionally fitted, either in large ships as an interim measure while port facilities are still being developed, or in small ships for self-sufficiency, when operating to the less well-equipped ports. In the latter case, containers are not usually landed onto the quay by the crane, but onto a trailer which is towed across the hatch covers to a stern ramp and down to the quay—the 'Strider/Tarros' concept. Hatch covers are very closely spaced (about 90 mm apart) so that wheeled vehicles can readily drive across, as shown in Fig. 8.6.

The selection of access equipment for all but the smallest cellular container ships is more or less standardized. Each block of container stacks is covered by a pontoon cover, which is lifted off by crane whose spreader locks onto four stacking points. It is then stowed on an adjacent hatch cover or on the quay. Each cover must be within the lift capability of the available cranes,

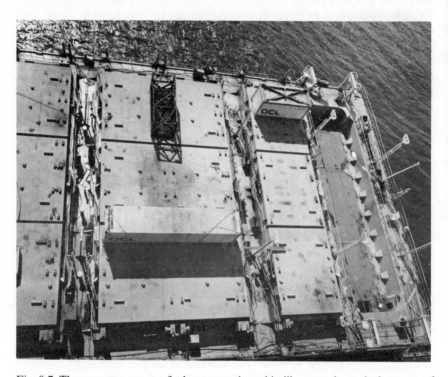

Fig. 8.7. The pontoon covers of a large container ship illustrate the typical pattern of stacking points. Each large cover can stow 40 ft containers (as illustrated in the middle) or two 20 ft containers, using the intermediate supports. The angled sockets are lifting points, for use with either a 40 ft or 20 ft crane spreader. No separate lashing points are fitted, as twist locks are employed. Each pontoon covers three container cells abreast.

usually 30 tons corresponding to a fully loaded 40 ft container. Covers generally span only one 40 ft stack in the fore-and-aft direction, but are up to five rows wide. They are thus about 13·0 m long for a single 40 ft (12·2 m) stack. A length of 13·7 m is needed to cover two adjacent 20 ft stacks, owing to the intermediate cell guides. Transversely, hatchways are a multiple of about 2·7 m. Two or three covers abreast are usual. Thus a 9-across ship would have three covers each about 8·0 m wide, spaced as close together as the construction of the longitudinal support girders permit. Such covers would weigh about 20 tonnes each, and there would be up to 40 in a large ship. The cost of the covers typically amounts to about 2% of the ship cost.

Handling a large number of covers is no problem in a big ship served by two or three shore cranes. In small ships, on short sea routes with a very quick turnround time, it is more common to fit ship-operated hatch covers for speedier operation, such as multi-panel end folding covers or hydraulically operated lift and roll covers.

In all container ships hatch covers are fitted with stacking points for deck containers. Often fittings for both 20 ft and 40 ft containers are provided on each cover (see Fig. 8.7), so that a variety of different container mixes can be accommodated without having to move cell guides in the holds. Covers are usually only designed to support containers of 12–15 tonnes average weight per TEU, so that if three or four tiers are to be carried, the upper tiers must normally consist of either light containers, or empties (2–3 tonnes). The design of covers and lashing arrangements are described in Sections 4.5.5, 5.9 and 9.1.

Arrangements for other than 20 ft and 40 ft containers are similar but such containers are rarely used, except by a few operators employing both their own containers and their own ships, e.g. 35 ft (10·67 m) by Sea–Land, 30 ft (9.12 m) by British Rail, 24 ft (7·32 m) by Matson, and 10 ft (2·99 m) by DFDS.

8.2.2 Container ship variants

Although almost any type of dry cargo ship can be adapted to carry containers, including bulk carriers, two variants are worth mentioning. The hybrid ship can combine container capacity with general cargo and/or Ro-Ro cargo capacity. In certain ships the container spaces, both on and below decks, are arranged as described above, but only part of the total space available is given over to this purpose. The remainder is arranged for Ro-Ro cargo as described in Section 8.3.3.2.

Feeder container ships are usually built as small cellular ships, but a few designs carrying only deck-stowed containers handled by crane or over a ramp have been developed. No cargo access equipment is needed for the main hull, as it only accommodates machinery, tankage etc.

8.3 Roll-on/roll-off ships

An extremely wide variety of ships is included in the category of roll-on/roll-off vessels. The principal types are described below, together with some considerations influencing the selection of their access equipment. Although typical examples are given, it must be emphasized that a wide variety of combinations of external and internal access equipment specifications is available; early discussions between ship operator, shipbuilder, port authority and equipment manufacturer will assist in finding the best solution.

8.3.1 Passenger/vehicle ferries

The first roll-on/roll-off vessels for road vehicles were small ferries for cross-estuarial or island services. These carried a few wagons or cars in addition to their normal passengers. Vehicles were carried on the open deck, and driven on or off by simple ramps, or even a few planks on to a quay or slipway. On longer routes, the few vehicles requiring passage were lifted into conventional holds by derrick or crane. By the 1940s, a few pioneering designs for larger passenger ferries made special provision for motor vehicles, giving over part of one deck to their carriage, with access over the open stern via a simple ramp.

The designs evolved during the 1950s, with North European short sea ferries the pacemakers. Typically these had a completely enclosed vehicle deck extending over virtually the entire length of the ship at freeboard deck level, i.e. the first complete deck above the waterline. Access was via a stern door, which also served as a connecting ramp to an adjustable bridge ramp ashore (linkspan). The vehicles carried were mostly tourist cars, although the after part of the deck was strengthened for commercial vehicles. In many cases about 100–150 cars could be carried, plus about 1000 passengers.

With the growth of private car ownership, and the increasing amount of trade across the North Sea in trucks and trailers, the passenger car ferry design was refined to increase its size and its flexibility to cater for a varying seasonal mix of commercial and private vehicles. Such vessels are nowadays usually operated by long-established companies and serve regular routes with purpose-built terminals. They can therefore be closely tailored to suit a particular trade. A typical modern design of passenger/vehicle ferry is shown in Fig. 8.8 which exhibits many of the following common features of such vessels:

  (i) main vehicle deck is the freeboard and the bulkhead deck. No vehicles below this deck, only accommodation, machinery, stores, tankage etc.;
 (ii) generally both bow and stern doors to avoid turning vehicles, thereby speeding up the loading and discharging of passenger-driven cars. The ship and the terminal are arranged for both bow first and stern first operation;
(iii) slight sheer on deck to assist meeting subdivision requirements for passenger vessels (without sheer, flooding of a compartment near either end could soon submerge the margin line, 76 mm below the freeboard deck);
(iv) vehicle deck space exempt from GRT and NRT, being above the

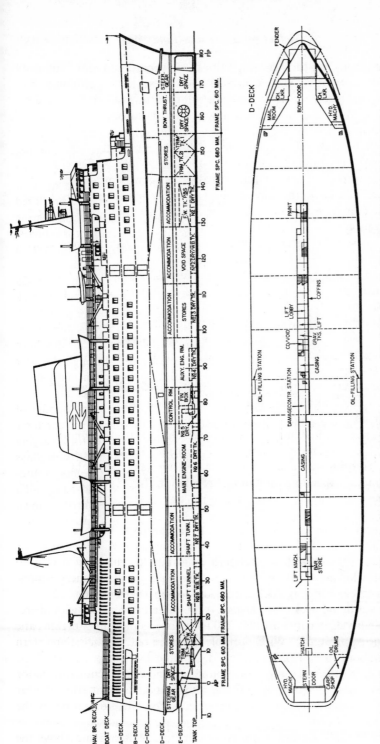

Fig. 8.8. This plan shows the profile and main vehicle deck of the passenger/vehicle ferry *St Columba*. This 7800 GRT 19½-knot vessel can carry 2400 passengers and 335 cars across the Irish Sea. The stern door is 5·5 m wide and 4·5 m high, the bow door 4·5 m square. Portable car decks provide 1400 m² for 150 of the cars; when they are raised, 36 lorries can be carried.

tonnage deck (in these ships also the freeboard and bulkhead deck);

(v) clear height of vehicle deck 4·3–4·5 m so that high commercial vehicles can be carried, and to permit the fitting of portable car decks to increase car capacity during the tourist season;

(vi) portable car decks fitted, with hoistable ramps at each end, which can be lowered to give about 1·9 m clear height, sufficient for tourist cars and leaving about 2·4 m clear below, sufficient for vans or camping vehicles. Decks, usually one or two lanes wide, but may well be full width, provided next to the casing for passenger access to stairways. Car decks may be hoistable, folding or pivoting (see Section 6.8);

(vii) vehicle deck arranged in lanes, 2·8–3·0 m wide for commercial vehicles, or 2·2–2·3 m wide for cars;

(viii) narrow casings for machinery uptakes, stairways, ventilation equipment etc. Usually on centre-line, or just off-centre, providing structural strength;

(ix) straight axial stern ramp, often doubling as stern door. Door dimensions vary for individual ships, from about 5 m wide × 4·5 m high in smaller ferries, up to 9 m × 5 m in larger vessels permitting two-lane operation;

(x) bow access is usually at least 3 m wide × 4·5 m high, closed by bow visor and internal ramp/door. Its dimensions are often limited by the shape of the ship's forebody;

(xi) typical axle loadings for the main vehicle deck are 16 tonnes on single axle or 20 tonnes per double axle spaced at least 1·3 m apart. These correspond to a maximum road-going vehicle weight of 35–40 tonnes, depending on national regulations (see Table 9.4). Portable car deck loading about 1 tonne per axle (2 tonnes per vehicle).

The typical North European or Mediterranean ferry built to this basic design is a ship of 19–24 knots, 130–160 m long overall, 20–24 m beam, 4·0–6·4 m draft and with a freeboard amidships to the vehicle deck of 1·5–2·4 m. Allowing for deck structure, the vehicle deck is from 4·9–5·2 m high beam-to-beam, giving a depth from keel to upper deck of about 12·5–13·5 m. Three additional decks for passengers are often fitted above without impairing stability. Capacities are 800–2000 passengers, and 250–400 cars (with portable decks lowered). With portable decks raised, commercial vehicle capacity is about 40–50 12 m trailers per deck, typically arranged in three lanes both sides of the engine casing. Corresponding deadweight is modest, usually between 1500 and 3500 tonnes, so that draft variation in service is small. The ship-to-shore ramp arrangement compensates for any tidal variation; in enclosed docks, therefore, very simple facilities are adequate. Routes may be as short as 20 miles (one way), but 100–500 miles are more typical. Medium speed diesels geared to twin screws are the most popular machinery, as they can readily be installed in the low headroom below the vehicle deck.

Modifications to the basic design include the following:

(i) the bow door may be omitted in larger vessels wide enough to permit vehicles to turn readily (and since they are on longer routes they can tolerate longer port time than smaller vessels);

(ii) beamier vessels stable enough for two main vehicle decks, employing either internal ramps, or two-level loading. Sometimes they have double-width stern doors;

(iii) where the deck above the vehicle deck is not also the forecastle deck, side-hinging bow doors may be installed to avoid the excessive weight of a bow visor (see Fig. 6.28);

(iv) casings may be fitted at the sides rather than on the centre-line. Occasionally there are twin casings which permit three main flows of traffic;

(v) large side doors, about 5 m square, may be fitted in vessels carrying cargo on the vehicle deck during the off-season, to allow two-way traffic onto conventional quays;

(vi) space for export cars can be provided below the freeboard deck, served by an elevator with watertight cover;

(vii) smaller garages can be provided on the upper deck, served either by an internal ramp from the main vehicle deck, or direct from the shore bridge ramp;

(viii) additional strengthening of a limited area of the vehicle deck may be provided for carrying very heavy loads, including military vehicles;

(ix) local conditions may permit special features, e.g. in Japanese inter-island ferries, angled bow and stern quarter ramps are common, and clear deck height can be 4·0–4·2 m, reflecting lower highway limits compared with Europe or America (see Table 9.4).

Cargo access equipment, including bow and stern doors and ramps and portable car decks, typically forms about 4–6% of the ship's cost. The percentage is not as large as in freight Ro-Ro vessels owing to the high cost of the passenger accommodation.

In addition to the 300 or so large passenger/vehicle ferries worldwide, there are also over 1000 smaller ones providing local services to islands or across estuaries. Most of these operate in comparatively sheltered waters, and vehicles are often accommodated on a single open deck. Some of the features discussed earlier still apply e.g. lane widths and axle loadings. Ramp equipment is usually very simple, often little more than a flap which accommodates tidal variations and lowers onto a sloping slipway as in Fig. 6.30.

8.3.2 Short sea freight roll-on/roll-off ships

8.3.2.1 *General considerations*

The first significant roll-on/roll-off vessels intended primarily for the carriage of commercial vehicles and freight were converted wartime tank landing ships operating from the U.K. to Ireland and the Continent. In the 1950s, newbuilding designs were developed which were in many ways similar in concept to the passenger/vehicle ferries described in Section 8.3.1 but without their extensive passenger accommodation. The main vehicle deck extended over almost the full length of the ship, with access via an axial stern ramp which also formed a watertight door. Additional cargo could be carried on the open upper deck aft of the superstructure, handled by ship or shore crane.

With the growth of road networks, particularly within Europe, manufactured goods increasingly moved door-to-door in road vehicles – rigid and articulated lorries and tractor–trailer combinations. Improvements in documentation and customs procedures assisted this trend. Throughout the 1960s the freight Ro-Ro continued to evolve, especially in the highly competitive North Sea area, as illustrated in Fig. 8.9.

The essential design problem is to ensure that the benefits of rapid port turnround time are not lost by restricted payloads due to the inherently low capacity of wheeled vehicles carried in a ship. For example, a large 12 m (40 ft) trailer requires a block within the ship's structure about 12·5 m long, 3·0 m wide (allowing for clearance and lashings) and about 5·0 m high (including the deck structure) – a volume of 190 m³. The trailer itself has an internal capacity of about 70 m³ in its body above the chassis, so only about one-third of the ship's potential capacity is readily usable.

Thus ship designs have evolved which make maximum use of all deck levels in the hull, compatible with reasonable access. Three decks are the norm except in the largest ships, typically:

   (i) the main vehicle deck, which is also the freeboard deck, only a few inches above the deepest waterline (although often raised locally at stern door threshold);

  (ii) the lower deck, usually the tank top, i.e. inner bottom of the ship. Since more than 12 passengers are not usually carried, close subdivision is not required. Access can be arranged by ramp or elevator to a space often extending to about half the ship's length. Accommodation for additional lorry drivers is sometimes provided, when some relaxations in regulations may be permitted compared with regular passenger vessels;

  (iii) the upper deck, usually the weather deck. The superstructure is small, as it is only required to provide accommodation for the crew (about 20 men), and operational spaces, e.g. wheel-house. It can be arranged

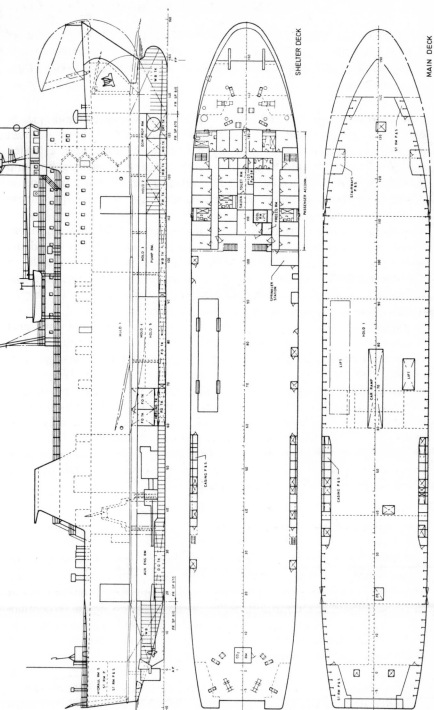

SHELTER DECK

MAIN DECK

Fig. 8.9. The plan shows a North Sea freight Ro-Ro, the 3500-tdw 18-knot *Duke of Yorkshire*. Her decks provide 3020 m² or 11 200 m³ bale capacity, equivalent to 180 TEU, or 52 trailers plus 60 cars. A 9 m long axial stern ramp serves the stern door, which is 7 m wide and 5·9 m high. The bow ramp is 4 m wide. Access to the upper deck (shelter deck) is by elevator, but by ramp to the lower decks. Owing to the subdivision requirements resulting from the accommodation for 90 passengers, the lower decks are restricted, and are primarily intended for cars.

in various ways to increase usable deck areas. A block forward gives protection to cargo stowed on deck, or there may be narrow casings aft with accommodation spanning the driveway between them. Alternatively a light structure may be built over a partly enclosed upper deck, as in Fig. 8.9.

Such arrangements have worked well in the rapidly increasing number of vessels built to serve routes all over Northern Europe. Ports have provided terminal facilities consisting essentially of a berth with a bridge ramp which can be adjusted for tide height and ship's access level, marshalling and parking facilities for vehicles, and administrative and reception buildings.

The majority of the vehicles carried consist of rigid or articulated lorries, road trailers and roll trailers loaded with cargo, e.g. containers, at the terminal and pulled by tractors, all usually driven by the stevedores rather than their road drivers. Although a few busy short sea routes have adopted small cellular container ships, the majority have chosen Ro-Ro vessels for their greater flexibility for all types of cargo, and simpler port facilities. Ro-Ro ships can accept containers either on roll trailers, or on large fork lift trucks. Some ships have been arranged to carry lightly-loaded or empty containers on the upper deck, handled either by ship or shore crane. This basic pattern has been refined further for larger and more advanced ships which, although designed primarily for short sea routes, have been operated on longer routes when demand has arisen, e.g. to the Middle East.

The majority of modern short sea ships are 100–140 m overall length, with breadths of 18–23 m, maximum drafts of 5–7 m and speeds of 16–20 knots. Their carrying capacity can be expressed in a number of ways, none of which is in itself a complete measure of their capability:

  (i) maximum deadweight (typically 3000–8000 tonnes);
 (ii) bale capacity (excludes open deck spaces);
(iii) linear lane length (total length of vehicle lanes 2·8–3·0 m wide on all decks suitable for commercial vehicles);
 (iv) trailer capacity (generally 12 m type);
  (v) container TEUs (Twenty foot Equivalent Units). Other units may be converted, e.g. one 15 m lorry = 2·5 TEU;
 (vi) number of particular units, e.g. cars.

The design of the ship may be arranged to increase one particular form of capacity, without affecting another. For example a deck 6 m high rather than 4·5 m will double the container capacity of that deck, but not change the lane length or number of trailers.

Thus a typical vessel might have the alternative capacity descriptions given in Table 8.1.

Even after allowing for the weight of some hundreds of tonnes of oil fuel, water and stores in the deadweight, such a vessel can easily carry a full cargo

Table 8.1. Typical short-sea freight Ro-Ro alternative capacities

	Lower deck (4·5 m high)	Main deck (6·0 m high)	Upper deck (mostly open)	Total
Maximum deadweight, tonnes	—	—	—	5 500
Bale cubic capacity, m³	3 000	10 500	—	13 500
Linear lane length, m	230	570	500	1 300
Number of 12 m trailers	16	44	36	96
(capacity at 70 m³ each)				(6 700)
Number of TEUs	32	176	144	352
(capacity at 30 m³ each)		2 high	2 high	(10 600)
Number of cars (including cars stowed on portable decks)	120	330	150	600

of 30-tonne 12 m trailers (2880 tonnes), or 1-tonne cars (600 tonnes). 352 fully loaded 20·3-tonne containers (7150 tonnes total) could not be carried, but it is most unlikely that a load consisting solely of 20-foot containers would ever be carried in a Ro-Ro ship. A full load of average weight 14-tonne containers (4920 tonnes total) is, however, within the ship's deadweight capacity.

Deadweight and draft rarely limit a Ro-Ro ship's capacity unless a lot of water ballast is required for stability or trim purposes. Careful consideration needs to be given to distribution of cargo, fuel and ballast, not only to assist cargo operations, but also to give a satisfactory metacentric height giving easy rolling motion with adequate stability.

The world fleet of small freight Ro-Ros has been expanding rapidly, and now numbers about 400 purpose-built vessels. Many ships have been ordered since 1974 as a result of the Ro-Ro ship's congestion-beating capability in booming Middle Eastern and West African ports. A number of converted or hybrid vessels with partial Ro-Ro capability also exist, but they have less elaborate access equipment. Conversions can be a means of obtaining Ro-Ro capacity more quickly than by newbuilding, (see Section 10.8) but they are not usually as competitive on short routes as purpose-built vessels.

Ro-Ro cargo handling rates are impressive compared with break-bulk ships. Single lane working typically averages 15–25 large vehicle movements per hour, allowing for marshalling, manœuvring and lashing. This corresponds to some 300–500 tonnes per hour. Higher rates up to 1000 tonnes per hour can be achieved for heavy regular cargoes like forest products – pulp, newsprint or paper – with two-way operation on the main deck and ramp. In short sea Ro-Ros, two lane operation allows empty vehicles, e.g. terminal tractors or fork lift trucks, to return more quickly for a new load. But discharging and loading are not often carried out simultaneously, unless the terminal is equipped for two level ramps, as operations on board become too complicated. Since port times are as little as 6–12 hours, short-sea Ro-Ros are hard worked ships requiring high standards of equipment reliability.

8.3.2.2 *External access equipment in small Ro-Ros*

The possible choices of external Ro-Ro access equipment depend largely on the geometry of the ship and shore arrangements related to the cargo loads moving. As discussed in Section 6.3.2, there is a maximum practical slope which each type of wheeled equipment can readily negotiate. The slope of any ramp linking ship and quay depends on its length and the relative heights of quay and ship. This must take account of tidal range and ship access level above the waterline at likely drafts and trims. In theory, the slope should be acceptable at the two extremes of deepest loaded ship and lowest tide, and lightest loaded ship and highest tide. In practice, a smaller range can be allowed for, since adjustments are possible to the ship, or to the bridge ramp if used, or simply by accepting short periods when extreme tides limit operations.

The simplest situation is where the height of the quay and vehicle deck are approximately equal, and there is little tidal variation, as in the Mediterranean or Baltic. External access equipment can then be a straight axial ramp which lies directly on the quay, with fingers to smooth the transition, as illustrated in Fig. 6.16. Variations in ship draft due to changes of load are not great in small Ro-Ro vessels, typically no more than 1–1·5 m. The after draft can be controlled within limits by transferring water ballast. In such cases, the shipboard ramp which bridges the gap can also serve as a bottom-hinged weathertight stern door. Thus it need have dimensions little larger than the access opening in the hull.

The height of opening depends on the clear height within the vehicle decks – about 4·5 m clear for road vehicles or 6·0 m for containers stacked two-high on roll trailers. Minimum clear widths of opening are about 4 m for single lane traffic, or about 7 m for two lanes, although in practice 5 m and 8 m respectively are more common. Stern ramps which are wider than they are long are usually arranged in two sections, for increased structural flexibility when the ship heels. The load-carrying capability is determined by the heaviest loads anticipated in the ship. A typical specification suitable for a fork lift truck carrying 20 ft containers would be:

maximum front axle load 50–55 tonnes;
maximum loaded vehicle weight 65 tonnes.

A lighter loading is possible if roll trailers are used instead.

If the quay is not strong enough to withstand the weight of the ramp plus its load, chains can be deployed so that the ship takes the greater part of the load by cantilevering the ramp (see Fig. 6.19).

This form of stern ramp/door is also suitable for the increasing number of ports which have a bridge ramp, whose level can be varied with the tide as shown in Fig. 2.4, and having a suitable slope, width and load capacity for the ship. Where both terminal and ship are being built specifically for a particular trade, the design of each can be matched. Similarly, where either a terminal *or* a ship is being built, the other being already in service, compatibility can easily be assured. Where a ship regularly serves a route, the bridge ramp may accommodate all relative movement, so that the ship requires no stern ramp at all, only a weathertight door.

Ships having a stern door that must be opened before arrival at the berth are potentially dangerous as water may enter over the low threshold and flood the vehicle deck. The resulting large free surface endangers the stability of the vessel. Problems can arise where ships are built for routes having a variety of berths and terminals. While absolutely standard berths are unlikely to be available, some harmonization is possible, to reduce the number of mismatches. To this end recommendations have been proposed by the International Association of Ports and Harbours [8.1] covering such considerations as;

(i)   limits of slopes of shore ramps (fixed inclines on quays on which the ship ramp may rest, generally in ports with small variation in water level);

(ii) provision of bridge ramps or linkspans and recommended slopes, heights and widths;

(iii) pedestrian access.

For ships where a short stern ramp is insufficient to permit operation at all ports of call, longer ramps may be fitted to achieve more acceptable slopes and heights. If the ramp cannot be adapted, e.g. by folding, to serve as a door as well, a separate watertight door must be fitted. This usually folds upwards (see Section 6.3.9). A separate door is also often fitted where very heavy loads are anticipated which might distort a ramp/door and reduce its watertightness. Ships so equipped can operate to many ports without bridge ramps by berthing stern on to a shore ramp or ordinary quay, either in a corner or at right angles to the quay. In the latter case (Mediterranean moor), the ship lets go its anchors to hold the bow in position as it manœuvres its stern to the quay (see Fig. 6.17). Very little quay space is required, which is valuable in congested ports, but ships mooring stern-to require fairly still water and must not be so long that they interfere with other port operations.

The smaller freight Ro-Ro can often be made versatile using only variations of the straight axial stern ramp. For example, a 12 m long ramp can accommodate a variation in height of +1·5 m to −1·5 m from stern door threshold with a slope not exceeding 1 in 8, with any further variation being taken up by a bridge ramp. Angles of heel of up to 4–5° can be readily accommodated, and trims of 2°. The angled quarter ramp provides versatility too, especially in larger ships, but it has a number of disadvantages which make it generally unsuitable for ships below about 6000 tdw, including:

   (i) greater weight and cost, including separate stern door;
  (ii) roadway narrower than stern door opening which may restrict two-way traffic;
 (iii) length 60–80% greater to give same vertical range at quayside, owing to relative geometry and critical distance (see Fig. 6.12);
 (iv) off-centre load of ramp and unsymmetrical stowage of cargo in aft region can produce large angles of heel. For example, simply lowering a quarter ramp on to the quay can cause a heel of 5° in a small ship. Heel during cargo operations can introduce problems with the ramp bearing unevenly on the quay;
  (v) loss of stowage space in the aft region owing to position of door and manœuvring space for turning vehicles in a narrow ship.

Furthermore, as more ports provide proper Ro-Ro facilities, either

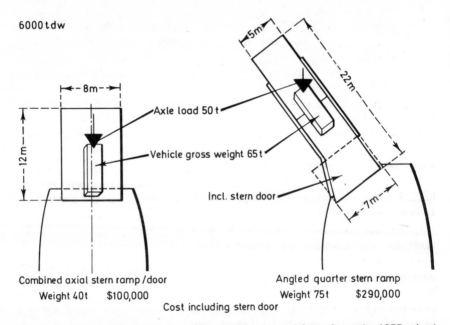

Fig. 8.10. This drawing [6.3] shows the comparative weight and cost (at 1977 prices) of an axial stern ramp and an angled quarter ramp in a 6000 tdw Ro-Ro.

permanent bridge ramps or floating linkspans, the need for versatility can be expected to diminish.

Fig. 8.10 compares a straight and an angled ramp in a 6000 tdw ship, as reported by Hanson in [6.3].

Other forms of external cargo access on short sea vessels, although less common than the single axial stern ramp and door, include:

(i) two level access, with a shore bridge ramp serving an upper deck as well as the main deck, either a two-tier ramp at a special terminal, or an adjustable bridge ramp to serve each deck as needed. A separate upper ship ramp is usually fitted to link with the bridge ramp;

(ii) two or three stern doors, to provide extra width or traffic separation with different doors serving different decks via internal ramps;

(iii) bow door and ramp as in passenger/vehicle ferries. With dimensions restricted by bow shape and the smaller number of vehicles on freight ferries, a drive-through facility offers no great advantage except in special cases, e.g. export cars, or multi-port discharge. For experienced drivers, manœuvring vehicles backwards into a ship for loading via a single stern access is no problem;

(iv) side doors, for use when stern ramp access is restricted;

(v) hatch cover in upper deck giving vertical access to the main vehicle deck; sometimes in the main deck as well, to allow a crane to plumb the lower hold. Early Ro-Ros often had a flush hatch, folding or pontoon, partly as an insurance against the possible failure of the stern ramp or door. Reliability of the latter has proved such that hatches are now only fitted if required for special purposes (unless covering an internal ramp from below – see Section 8.3.2.3).

It can be seen that a wide variety of configurations are possible. Where the trade requires it, special features can be designed into the Ro-Ro vessel more readily than almost any other type. The cost of the access equipment depends on the specification. A typical outfit costs between 6% and 8% of ship cost.

8.3.2.3 *Internal access equipment in small Ro-Ros*

Internal access to the various decks in a Ro-Ro must be arranged in such a way that the ship can be regarded as 'open' in the horizontal sense, in the same way that a cellular container ship is 'open' in the vertical sense.

The first requirement is for wide clear decks, as far as possible unencumbered by pillars or casings, to assist traffic circulation. Most short sea freight Ro-Ros have casings for access to the machinery, ventilation and exhausts. These may be located on one side of the ship or both, and are usually no more than 2 m wide. Usually a deck structure can be built to span the ship's breadth without intermediate supports. Decks over vehicle spaces are typically supported by a number of deep transverse beams supported at the ship's side by web frames, with deck plating and smaller stiffeners distributing the load onto the beams. The depth of this structure depends not only on its span, but also on the loading on the deck above, the spacing of the beams and the scantlings of their structure. As a first approximation, the depth of the beam averages about 5% of the span between supports (ship's side, casing or pillars) but can vary between about $3\frac{1}{2}$% (for lighter loads) up to about 6% (for wider-spaced beams without very large face flats). Thus in short sea Ro-Ros with breadths up to about 20 m, such beams rarely exceed 1 m in depth, especially if the deck above is cambered to increase its depth at mid-span, e.g. at a weather deck. Adding the beam depth to the clear height underneath gives an indication of the deck-to-deck height required. The ship designer needs such information early in the design, so that ship breadth and depth can be chosen to give satisfactory stability and stowage arrangements. A breadth of about 19–21 m is popular as it is suitable for six lanes of heavy vehicles 2·8–3·0 m wide, or for three 20 ft (6·06 m) containers stowed athwartships. As a first approximation, breadth should not be less than about 1·4–1·5 times the depth to the upper deck.

Although the main deck carries a large proportion of the cargo in a short sea Ro-Ro, it is still necessary to ensure that cargo can be readily transferred to other decks. Usually some form of internal access is required, even if two-level external access is possible. One means of access between decks is a sloping ramp, which permits continuous circulation of traffic. In practice, it is often difficult to arrange satisfactory internal ramps in small Ro-Ros because of slope limitations and manœuvring requirements. Elevators are likely to be more expensive and have the drawback of a discontinuous cycle, but they take up considerably less space in the ship.

Since Ro-Ro ships can readily be arranged to suit different trades and cargoes, it is always worth examining whether some ramps, particularly of the folding type, can be installed without great penalty. For instance, since the lower deck may only comprise about 15% of the trailer lane length, and is awkward and slow to stow, it may be possible to dispense with it altogether in passenger/vehicle ferries on short routes. Alternatively it may be used only for 'easy-to-stow' cargoes like export cars, tractors, palletized cargo and even bulk liquids.

Suitable vehicles can operate with low headroom, and they can negotiate steeper gradients so that the length of any ramps fitted need not be excessive, e.g. a deck-to-deck height of 2·3 m with a slope of 1 in 6 gives a ramp 14 m long. Space may be gained in critical areas by introducing local deviations like sloping recesses or bulges outside the normal ship's side.

Access for large vehicles to the upper deck is essential to provide a reasonable ship capability. A number of ramp solutions are possible; detailed descriptions are given in Section 6.3.10, but the main alternatives are:

(i) a fixed ramp, with an opening at the upper deck covered by a flush hatch or a trunked passage having a watertight door.

(ii) a movable ramp, which can stow in a number of positions: one end raised to the deckhead, or to form part of the deck in the closed position.

A fixed ramp is cheaper than a movable one, but it blocks off a greater amount of stowage space and so it is not often fitted in small Ro-Ros. It can however be given an S-shape profile to reduce the change of gradient at each deck, but with the steepest possible mid-section for the loads envisaged, say 1 in 5 for cars or 1 in 7 for heavier vehicles as indicated in Fig. 6.8. Minimum ramp widths vary between about 2·5 m for cars to 3·5–4 m for heavy vehicles. Ramps 7 m wide are required for 20 ft containers stowed athwartships, but are difficult to fit in small vessels. Fig. 8.11 [6.3] illustrates three alternative internal ramp arrangements with comparative costs. The weights shown are mainly of concern to the ship and equipment designer, since the differences in weight do not result in differences in cargo payload. In this type of ship payloads are determined largely by space considerations.

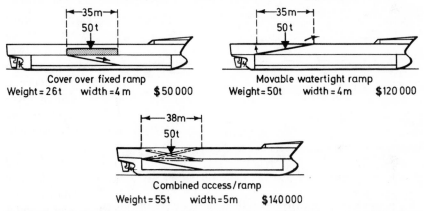

Fig. 8.11. This drawing [6.3] shows the comparative weight and cost (at 1977 prices) of a simple side-hinged cover over a fixed ramp, a movable watertight ramp/cover and a three-position multi-purpose ramp. In the first and third cases, the weight and cost of the fixed ramp to the tank top are not included; approximate figures would be about 25 tonnes and $20 000.

Portable car decks may also be fitted in freight Ro-Ros for export cars. They are usually in sections spanning the width of ship and are lowered from the deckhead as required, with an access ramp at the aft end. As there are no passengers with their cars, a tighter stow and slightly less headroom is possible, typically about 1·7 m clear.

In many small ships, the elevator is a better overall solution to the internal access problem as a variety of well-engineered designs taking up little space are available (see Section 6.5), capable of operating hundreds of cycles per day. The depth of the surrounding deck structure is usually less than for ramps, whose openings are longer. For access to the lower deck, a centre-line position at the aft end of the stowage space is usual, just forward of the engine room. The dimensions and capability of the elevator depend on the anticipated cargo – about 13 m long by 3·0 m wide with a capacity of 40 tonnes and an operating cycle of about 3 minutes its typical for 12 m road vehicles or roll trailers. A scissors type elevator is popular. This lowers into a recess in the double bottom. It can either stow in the main deck which loses one trailer space in the lower deck, or stow flush in the lowered position, with the main deck opening closed by a side-hinging cover.

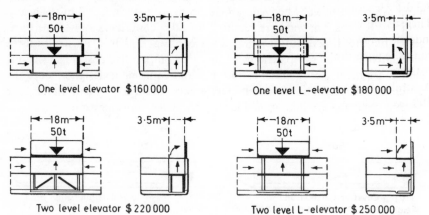

Fig. 8.12. This drawing [6.3] shows the comparative cost (at 1977 prices) of one- and two-level elevators, of the two-way and three-way (L-type) loading types, and their associated covers.

Elevators linking the main and upper decks are arranged in a variety of ways when they are fitted in preference to folding ramps. A position to one side of the ship is usually selected (although not so close that the ship's structure is not well supported over the length of the opening), in the after third of the length. An elevator capable of taking an 18 m lorry and trailer requires a platform about 19 m × 3·2 m with a capacity of at least 40 tonnes. A popular type is one which lowers down to lie on top of the main deck, with access to the platform by a short ramped section; it forms a watertight cover in the raised position (see Fig. 6.37). It is raised by wires or chains, usually actuated by hydraulic rams; chains give more positive control, need less maintenance and are considered safer.

If the elevator is not near the ship's side or other permanent structure, portable guides can sometimes be used. End-on loading of elevators is most common, but if three-way loading is required and permitted by safety regulations, e.g. for smaller vehicles to drive across the side, an L-shaped elevator can be installed, recessed into the deck. As Fig. 8.12 [6.3] shows, such a design adds about 13% to the cost of a two-way elevator. The figure also shows that a two-level lift can be provided for only about 38% more than a one-level lift, but such an arrangement requires greater control of traffic on each level, and may increase overall operating cycle time. Its position in the ship is limited and it requires equal heights for both decks for minimum cycle time. If cheaper elevators are required, cost can be reduced by reducing dimensions or capacity, e.g. so that only vehicles up to say 15 m long can be stowed on the upper deck, or by increasing operating times.

8.3.3 Deep sea roll-on/roll-off ships

8.3.3.1 *General characteristics*
There is no clear boundary between short sea and deep sea Ro-Ros, and indeed, many of the former have been chartered for use on long distance routes, e.g. to the Middle East.

Conversely some large ships are used on comparatively short routes, e.g. in Australasia and the U.S.A. Nevertheless, most of the world fleet of nearly 100 deep sea ships exhibit many of the following characteristics:

   (i) deadweight over about 12 000 tonnes;

   (ii) overall length over 170 m;

   (iii) at least four deck levels, sometimes as many as seven including car decks;

   (iv) a smaller proportion of road-going vehicles and a greater proportion of block-stowed cargo, i.e. cargo which does not travel on wheels on voyage, like forest products, pallets or containers (exceptions are the large ships used in U.S. trailer operations);

   (v) wide range of cargo stowage spaces, e.g. some permanent car decks, refrigerated chambers, bulk liquid tanks or cellular container stowage below decks;

   (vi) more extensive compartmentation below the main deck, usually double hull at sides, and more transverse bulkheads with watertight doors;

   (vii) more deck pillars to support greater spans;

   (viii) larger draft variation, owing to greater cargo load; maximum draft of about 10 m;

   (ix) may carry their own cargo handling equipment, e.g. large fork lift trucks;

   (x) no passenger capacity.

The larger hull dimensions of deep sea Ro-Ro ships not only permit a greater choice in internal layout, but also allow features less suitable for smaller vessels, e.g. fixed internal ramps or slow-speed direct-drive diesel machinery. Some owners prefer slow-speed to medium-speed diesels, and it is sometimes possible to accommodate this much taller engine without taking up too much cargo space, e.g. under fixed internal ramps. Machinery is always aft, with the engine casing usually arranged on or near to the centre-line.

8.3.3.2 *Ro-Ro/Container ships*

The first deep sea Ro-Ros were built in 1967 for ACL's North Atlantic routes. They have a deadweight of about 16 000 tonnes and serve developed ports with berths having both bridge ramps and container cranes. They were designed to carry Ro-Ro cargoes, including export cars, in the after part and containers in cells forward and stacked on the upper deck. Ro-Ro access equipment includes a stern door and axial ramp, and internal access is provided by fixed ramps. This basic design has since been refined in detail in later ships, as illustrated in Fig. 8.13, to cater for differing proportions of the three main types of cargo (Ro-Ro, cars and containers). The types of Ro-Ro unit vary considerably with the trade, but a rough overall long-term average might be one half containers in various sizes and the other half divided approximately equally between roll trailers, export cars, road trailers or lorries, rollable vehicles (e.g. construction equipment) and block-stowed cargo.

Typical requirements for such a vessel serving developed ports with a high percentage of container cargo would include the following:

(i) upper deck used for stowage of 20 ft and 40 ft containers up to four high, handled by shore crane, with no access from Ro-Ro decks below;

(ii) cellular holds forward for 20 ft and 40 ft containers stacked six to eight high;

(iii) stern access to main Ro-Ro deck with axial ramp at least 8 m clear width, to carry axle loads of at least 50 tonnes spaced 1·7 m apart;

(iv) stern door at least 6·5–7 m high to permit exceptional loads;

(v) a 'cathedral' area aft on the main vehicle deck for exceptional loads about 6·5–7 m high and 40–50 m long, with deck strengthened to permit distributed load of about 10 tonnes/m² (or equivalent in axle loading);

(vi) main vehicle deck 5·6–6·0 m high to permit two containers high block-stowed (i.e. lashed to deck after loading by fork lift truck or other vehicle), or three levels of export cars with two portable decks and clear height of 1·7 m;

(vii) second main vehicle deck 4·5 m high for trailers, served by a ramp at least 8 m wide with maximum slope of 1 in 10. Deck loading is typically 40 tonnes per axle, or about 3 tonnes/m² distributed;

(viii) fixed decks for export vehicles, mostly 1·7 m high, with some areas capable of accepting vehicles up to about 3 m high, e.g. light vans,

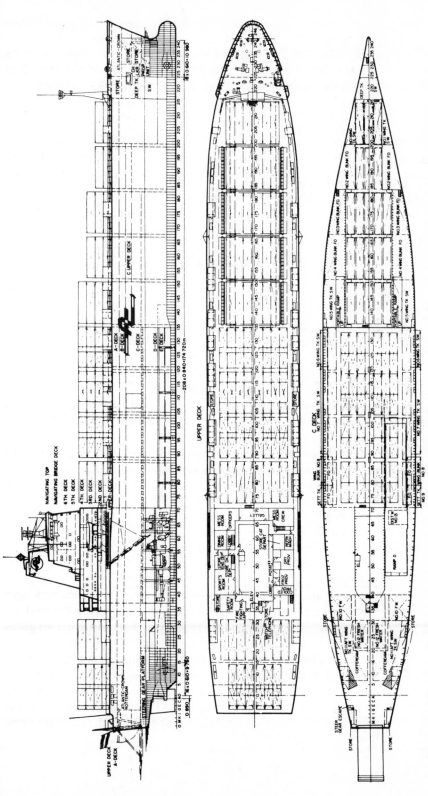

Fig. 8.13. This plan of ACL's *Atlantic Crown* shows the arrangement of a Ro-Ro/container ship. With a deadweight of 18 500 tonnes and a speed of 24 knots, she can carry about 800 TEU of containers in cells forward and on deck. Ro-Ro capacity aft and abreast the container cells consists of the equivalent of a further 150 TEU, plus 950 cars. A 15 m axial stern ramp and fixed internal ramps provide access to the six Ro-Ro levels. Container cells are also fitted, but rarely used in service. Container cells are covered by flush pontoon covers.

mini-buses, mobile homes, agricultural vehicles etc., served by ramps at least 2·8 m wide and having a maximum slope of 1 in 6. Deck loading is generally 1 tonne per axle, (about 0·25 tonnes/m²) except in the high area where loading is approximately doubled. Narrow spaces abreast cellular holds can also be used as car decks;

(ix) side doors, fitted in early ships for emergency access to the main vehicle deck and access to the car decks, are no longer necessary.

8.3.3.3 *The pure deep sea Ro-Ro*

The next major development in deep sea Ro-Ros was the Scan–Austral *Paralla* class ships of 1971 with a deadweight of about 20 000 tonnes, a bale capacity of about 50 000 m³, equivalent to about 1200 TEU and capable of operating to ports without special Ro-Ro facilities. The background to this Scandinavian development is well described elsewhere [8.2]. Access is by an angled quarter ramp which can be lowered on to more or less any conventional quay. This may be necessary either because there is no special Ro-Ro berth, or because it is already occupied. Since a shore container crane may not always be available, there are no container cells and all weather deck cargo must be handled via internal ramps. The first ships of this type were employed on the Scandinavia–Australia route, where there was a large amount of cargo which was not easily containerized, but which was suitable for handling by Ro-Ro methods – forest products, steel products and pallets. The ships are completely self-sustaining, carrying their own fleet of equipment to handle containers and trailers – typically six to eight fork lift trucks with capacities up to about 25 tonnes.

The concept has been further developed and employed on over a dozen routes. A typical vessel is illustrated in Fig. 8.14. The majority now exhibit the following broad characteristics:

(i) deadweight at least 15 000 tonnes, bale capacity at least 30 000 m³, over 800 TEU;

(ii) main vehicle deck 10·5–12 m above keel (i.e. draft of about 9–10 m plus freeboard of about 1–2 m);

(iii) access via angled quarter ramp, at least 7–8 m wide, capable of reaching quays situated between about 2 m above the stern door threshold and 4–5 m below it. Maximum slope does not exceed 1 in 8. Normally the ramp is in three sections with an overall length of about 35 m (see Section 6.3.4). It can support fork lift trucks with 20 ft containers, having a gross weight of about 60 tonnes and an axle load of 55 tonnes. 'Jumbo' ramps capable of two-way working have also been specified, with a width of 12 m and a length of 50 m. This means that three fork lift trucks spaced 15 m apart may be on each lane simultaneously – a total load of about 400 tonnes. Such a ramp

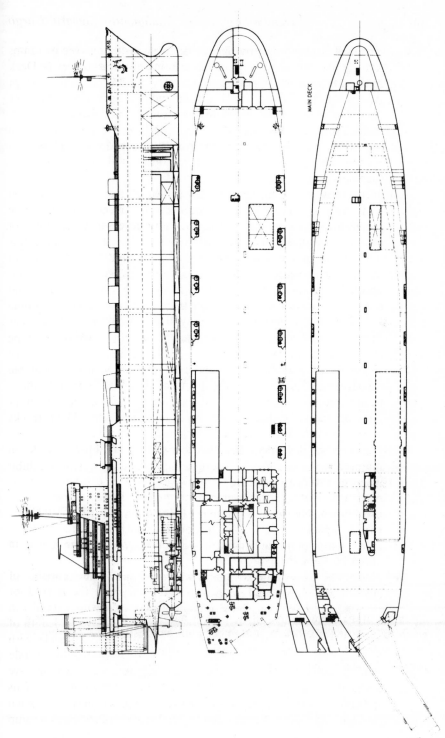

MAIN DECK

Fig. 8.14. The arrangement shows the modern 18 500 tdw 20-knot deep sea pure Ro-Ro *Skulptor Konenkov*. There are five decks providing about 10 600 m² of deck space, or 32 000 m³ of bale capacity, corresponding to about 750 TEU. A 36 m long angled quarter ramp provides the external access, with fixed internal ramps. The machinery comprises twin medium-speed diesels geared to a single shaft.

is unlikely to become a bottleneck, as it can accommodate handling rates of at least 60 units per hour, offering the potential of 24-hour turnround even in the largest ships;

(iv) stern door opening at least 11 m wide and 6 m high. The width corresponds to a broader inboard section of the ramp to assist manœuvring (see Fig. 6.12);

(v) low GRT and NRT owing to exemption of cargo spaces above the tonnage deck, although Suez and Panama Canal tonnages include such spaces;

(vi) decks above the main vehicle deck served by fixed ramps, at least 7 m wide to take a 20 ft container athwartships, with slopes preferably not exceeding 1 in 10, unless for access to car decks. The after ends of the ramps are fairly close to the stern door, so that traffic lanes rapidly separate from that congested area;

(vii) upper deck served by a ramp with a watertight trunk fitted with an upward-hinging door at least 7 m wide by 4 m high. Containers can be stowed on top of the trunk. An upward-hinging ramp which forms a watertight cover may also be used. While it is more expensive, and more difficult to keep watertight, it reduces loss of space to a minimum;

(viii) lower decks may be served by fixed ramp, movable ramps or elevators; the former being the most popular. Loss of space from a fixed ramp can often be reduced by starting the slope as far aft as possible, recessing slightly into the engine room area. The opening over a fixed ramp is covered by a side-hinging cover if the width is less than the clear deck height, otherwise an end-hinging cover is required;

(ix) where elevators are installed to save space, they are often arranged in pairs to provide a faster overall handling rate, and a reserve against breakdown. Dimensions are likely to be about 18–19 m by 3·2 m, with a capacity of 50 tonnes, but double-width elevators are available, capable of lifting two fully-loaded lorries plus trailers, a total of about 80 tonnes;

(x) specialized cargo compartments where required, e.g. refrigerated chambers or bulk liquid tanks.

Owing to the large number of containers moving deep sea, compared with road-going vehicles, efficient container handling is essential. Different parts of the ship are likely to be arranged for handling containers in the various forms in which they may arrive at the ship, for example:

(i) decks 6·0–6·2 m clear height for 8 ft 6 in (2·59 m) containers two-high on roll trailers or LUF units, or block-stowed two high by overhead-spreader fork lift trucks, C-vans or side loaders (sidelifts);

(ii) decks 5·4–5·6 m high for containers block-stowed two-high by bottom-loading or side-spreader fork lift truck (slightly less for 8 ft high containers);

(iii) decks 4·4–4·6 m high for containers one-high on road-going trailers, or block-stowed by overhead-spreader fork lift trucks, side-loaders or straddle carriers (see Sections 6.1.2 and 6.1.3);

(iv) decks 3·4–3·7 m high for containers one-high on roll trailers or LUF units, or block-stowed by bottom-loading or side-spreader fork lift trucks, or C-vans, or block-stowed products;

(v) decks 2·8–3·0 m high for other cargo on flats or roll trailers, block-stowed products, e.g. pallets two-high, or small commercial vehicles;

(vi) decks 1·7 m high for export cars.

Different operators favour different types of equipment, and carry different proportions of 20 ft and 40 ft containers, 8 ft or 8 ft 6 in high, or higher, and specify the ships' equipment to suit the trade. Handling of 40 ft containers on board ship is not possible by existing front-loading fork lift trucks, so that one of the other methods is required, e.g. roll trailers. Trades generating a large number of 40 ft containers are mostly those between developed countries, where shore cranes exist for handling deck-stowed containers.

Axle loadings for many of the vehicles listed are in the 40–50 tonne range; decks over about 4 m high are typically stiffened for 50–55 tonne axle loads, . those below, for 30–40 tonnes. Corresponding uniformly distributed loads are about 2·5–3·5 tonnes/m², with an area near the stern door strengthened for heavier indivisible loads, having a clear height of 7 m or so.

Advanced vessels of this type have the most extensive range of access equipment, which can amount to 8–10% of ship cost.

8.3.3.4 *Unnecessary over-specification can be expensive*

It is very easy to spend large sums, amounting to several million dollars, making a Ro-Ro ship as versatile as possible. Excessive expenditure can be reduced in a number of ways:

(i) careful analysis of the expected range of traffic moving, so that load-carrying capabilities, deck heights etc. are not too generous over the entire ship;

(ii) restriction of the largest and heaviest loads to the main vehicle deck. If other decks can cope with the less demanding traffic, benefits can be obtained like increased ramp slopes, and reduced deck loadings;

(iii) possibility of equipping only one ship in a fleet with an extra capability. Loads requiring its full capability may be rare enough for it to be possible to schedule the ship accordingly;

(iv) limiting the ability to handle every item of cargo every hour of the day in almost any port in the world. Extreme tides may occur only a few times in a year, ballasting may be possible, or such ports visited infrequently, and then not with the most demanding cargoes. Since stern ramp costs rise directly with their length, a reduction in tidal range to be catered for can permit shorter, cheaper ramps, normally without seriously affecting operational performance. An analysis of the ports and berths likely to be served must be made before finalizing a specification; cargo access equipment manufacturers are able to advise potential clients;

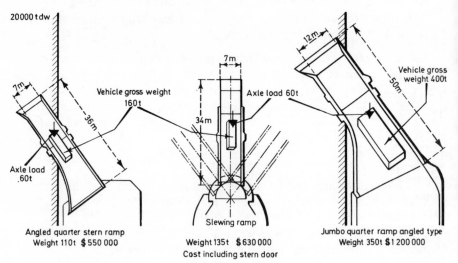

Fig. 8.15. This drawing [6.3] shows the comparative weight and cost (at 1977 prices) of three types of stern ramp in a 20 000 tdw Ro-Ro.

(v) examining whether the most sophisticated equipment is essential. For example, a slewing ramp enables a ship to berth on either side to or stern on. Fig. 8.15 shows that its cost is 15% more than an equivalent angled ramp. The narrow bottleneck section of the ramp is longer. It is usually necessary for engine casings to be at the sides to provide sufficient manoeuvring space;

(vi) investigate whether an existing design can be used, although it must be admitted that there are no great economies of scale in ship or equipment construction; the principal advantage of using 'standard' designs is a saving of time;

(vii) reviewing detailed specifications of access equipment. Additional expenditure to reduce the cycle time of equipment in more or less continuous operation like elevators has a much higher pay-off than expenditure on equipment operated only once or twice per port call, like stern ramps;

(viii) questioning the permanence of port limitations. Ro-Ro operation is here to stay, so ports with inadequate berths will not flourish unless they improve facilities. For example, the design quay pressure exerted by angled ramp is usually restricted to the low figure of 2 tonnes/m². This increases the cost of the ramp, compared with that of one exerting 3–5 tonnes/m² which many modern berths can accept;

(ix) examining whether exceptional loads can be carried by distributing them over a large number of comparatively lightly-loaded axles.

The operator who takes the trouble to make investigations into alternative configurations at the preliminary design stage is likely to end up with a more satisfactory product, more closely suited to his requirements.

8.3.4 Other Ro-Ro vessels

8.3.4.1 *The pure car carrier*

The 1960s also saw the introduction of the pure car carrier, to carry the large number of export cars being produced in Western Europe and Japan for shipment to, particularly, the United States, and returning in ballast. There are now about 100 specialized ships in the world fleet, capable of carrying about two million cars in a year. The basic design is simply tiers of garage decks of maximum area and minimum height. Cars are driven aboard via ramps and side doors. Early ships could carry about 2000 cars on eight decks, but more recent designs can carry about 5000 cars on 12 decks. Conventional measures of ship capacity, like deadweight, are of little significance for such light cargoes. Depending on its size 8–8·5 m² of deck area or about 4·4 m lane length 1·9 m wide are required for each car. This corresponds to about 40 000 m² area or 20 km length for a full cargo in the larger ships.

Fig. 8.16 shows a typical vessel with most of its decks having only 1·65 m clear height (1·90 m deck-to-deck). Most transverse bulkheads stop at the freeboard deck. Where fitted in the cargo spaces, bulkheads have watertight doors below the freeboard deck. The freeboard deck, often the fourth deck above the tank top, usually has a greater clear height to accommodate commercial vehicles. It can sometimes be subdivided by a hoistable car deck, and has a greater load capability, typically 1·5–2 tonnes/m² compared with 0.18 tonnes/m² corresponding to axle loads of about 0·8 tonnes on the car decks.

External access is by means of side doors two or three decks high, usually two each side, with sills at the freeboard deck, which is typically 1–2 m above the waterline on a maximum draft of only 7·5–8·5 m. A vertically stowing ramp can be lowered to serve any of the decks covered by the door, with a slope not exceeding 1 in 6. The ramp usually lies at right angles to the ship, but on narrow quays, arrangements can be made for it to be angled to run

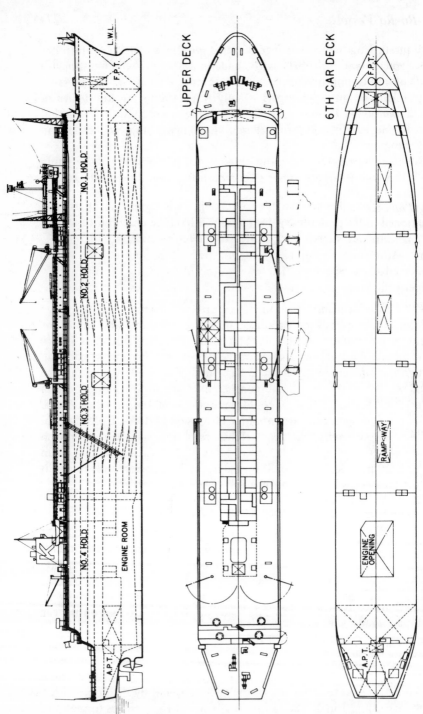

Fig. 8.16. This plan of *Toyota Maru No. 15* shows a typical pure car carrier with ten decks, with a clear tween-deck height of only 1·65 m. She can carry about 2800 cars, loaded and discharged via two side doors and ramps. The ramps are handled by the cranes shown, and stowed on top of the central deck house. Internal access is by a series of fixed ramps. Deadweight is only 11 000 tonnes, speed 21½ knots.

UPPER DECK

6TH CAR DECK

NO. 1 HOLD

NO. 2 HOLD

NO. 3 HOLD

NO. 4 HOLD

ENGINE ROOM

L.W.L.

F.P.T.

A.P.T.

RAMP-WAY

ENGINE OPENING

F.P.T.

A.P.T.

forward or aft. Internally, each deck is connected by two or three ramps two car lanes wide (about 4 m), which are usually fixed with an S-shape profile, although straight hinged ramps have been fitted. Cargo handling rates are high, over 100 cars per hour, so that even the largest ships can load or discharge in one or two days.

As in Ro-Ro vessels generally, safety precautions are extensive, with fire-fighting equipment, including CO_2 flooding, generous ventilation arrangements and lashing systems consisting usually of chains permanently fixed to the deck to which cars may be attached.

8.3.4.2 *Train ferries*
As mentioned in Section 2.6.6, train ferries were the earliest type of Ro-Ro vessel. There are nearly 200 in the world, linking mainland and island railway systems. Most train ferries carry freight wagons more frequently than passenger coaches. The typical ferry is basically similar in layout to the passenger/vehicle ferries described in Section 8.3.1: the main train deck is the freeboard deck, with machinery below and accommodation above. Indeed some ships are passenger/vehicle ferries with rails added at the vehicle deck, either recessed into the steel deck, or flushed off with wood or other material.

There is no cargo access equipment in the simplest train ferries, as the train deck is open at the stern, and a bridge ramp ashore connects with the ship so that wagons are shunted directly on board. For ships operating in exposed waters, a stern door is fitted. Those on short routes often have bow access as well. Four tracks abreast are common, with perhaps 400–500 m of track allowing 30–50 wagons to be carried. Since ferries are designed to operate on specific routes, rail gauges and clearances are selected according to the local standards.

Most train ferries operate on short routes of less than 100 nautical miles, so rapid transfer is more important than maximum carrying capacity. On longer routes, it may be worth utilizing lower and upper decks to carry additional wagons. Fig. 8.17 shows the ferry *Railship I* which carries up to 60 wagons up to 27 m long between Germany and Finland. It has 1300 m of track on three decks, served by an 80-tonne two-level elevator.

8.3.4.3 *Miscellaneous Ro-Ro ships*
The growth in Ro-Ro trades has also encouraged the development of other vessel types. Trailer-carrying barges towed by a tug have been used to the Middle East as described in Section 11.3.1. While less versatile than a Ro-Ro ship, a large capacity is obtained comparatively cheaply.

Vessels developed from the tank landing ship are used for carrying rollable cargoes which can be discharged directly via a bow ramp onto a beach through side-opening bow doors. These are often heavy lifts being delivered to remote construction sites without proper port facilities.

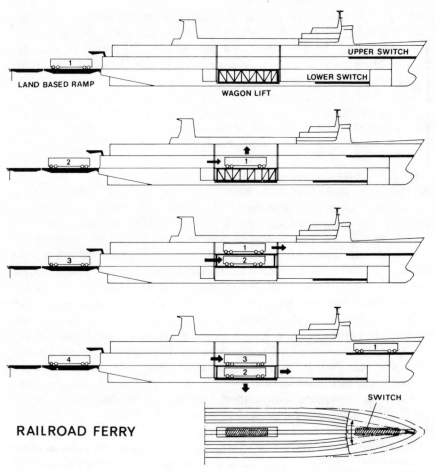

RAILROAD FERRY

Fig. 8.17. Most train ferries carry wagons on only the main deck, but this illustration shows how three decks can be used by means of a two-level lift and the corresponding loading sequence. The ferry is the 7000 tdw *Railship I*, operating between Travemünde and Hanko, which can carry sixty 20 m wagons at a speed of 20 knots. There are similar ships for operation in the Black Sea.

Other vessels with a Ro-Ro capability include hybrid multi-deck ships (Fig. 8.3), the Strider/Tarros type (Fig. 8.6) and some heavy lift ships (Fig. 8.21).

8.4 Specialized cargo vessels

8.4.1 Refrigerated ships (Reefers)

Refrigerated ships have been in service for nearly a century; demand continues for specialized ships for the worldwide fruit, meat, fish and dairy products trades, the banana trade being the largest.

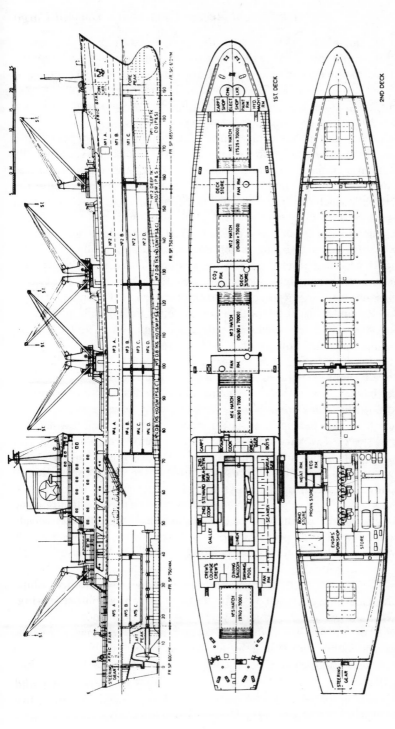

Fig. 8.18. A typical large reefer is shown, the 10 800 tdw 23-knot *Afric Star* with a capacity of 13 500 m³ (475 000 ft³). Four decks are fitted, dividing the ship into 18 compartments, whose temperature can range from +13° C to −25° C. Weather deck hatch covers are electro-hydraulic end-folding, and the tween-deck covers wire-operated folding, each with two banana elevator hatches.

Most of the world's 300 or so fully refrigerated ships are operated, basically as tramp ships serving the many seasonal trades, by a small number of specialist shipowners and charterers. Cargo liners with a small proportion of insulated capacity are built less often than before the advent of container ships, although some of the latter have a number of insulated holds (see Section 8.2.1), or carry refrigerated containers on deck.

The modern 'reefer' has evolved into a fairly uniform configuration: a four-deck ship (occasionally three decks) with either four holds (occasionally three) forward of the bridge and machinery space, and one hold aft, as shown in Fig. 8.18. The multi-deck arrangement not only provides many compartments which can be maintained at different temperatures (e.g. meat requires about $-25°C$, bananas $+12°C$), but also limits stacking height to avoid crushing cargoes such as cartons of fruit. Tween-decks are typically 2·6–2·9 m high beam-to-beam, which decreases to about 2·0–2·3 m after allowing for deck structure, insulation and air ducts. This height is compatible with the standard 1·8 m high pallet widely used in the fruit trades. The typical reefer has an insulated cargo cubic capacity between 10 000 and 17 000 m³ bale (350 000 and 600 000 ft³) with a corresponding maximum deadweight of about 8000 to 12 000 tonnes at a draft of about 9·2 m (30 ft). Light cargoes, like bananas, stowing at 3·3–4·0 m³/tonne (120–140 ft³/ton) give rise to the 'banana draft' condition for such ships of 6·1–7·6 m (20–25 ft) at which deadweight is reduced to about 4000–8000 tonnes. Reefer ships generally have speeds in the 20–23 knot range. Operation at banana drafts can add a further knot or so.

The cargo access openings must be arranged to minimize heat transfer into the cargo spaces. Insulated covers, of the minimum size compatible with efficient cargo handling, are required both at the weather deck and the two or three tween-decks. Side doors are usually fitted in the upper tween-deck to assist loading cargoes from the quayside by fork lift truck, or from lighters, and discharging via conveyors. Such doors can also be used if cars are being carried on what would otherwise be ballast passages. Typical side doors are 2 m square, one per side per hold. They are insulated and are either in one piece with special hinges so that they open outwards, flush against the hull, plating outwards, or they are simple hinged pairs folding back against the hull.

Hatchway sizes are restricted to about 30–35% of the ship's breadth, i.e. about 6·0–7·5 m wide, with lengths typically 45–55% of the hold length, i.e. about 9–12 m long for the midship holds. The comparatively small size of the hatchways limits the direct sunlight into the cargo spaces during loading in hot countries. Partial opening of the covers is desirable so that the minimum area is open to the air. Weather deck covers may be arranged so that, for example, only one panel of a folding pair need be opened; lift and roll covers are also used. Very often, two separate 'banana elevator' hatches about

2·5 m × 2 m are fitted in the corners of the main hatch, with a separate opening mechanism. The banana elevator may be lowered through the series of hatches to serve any tween-deck. Rapid operation of covers is necessary to reduce damage from rain. The tween-deck hatchways are the same size as the weather deck or slightly smaller. Wire-operated folding pairs are commonly

Fig. 8.19. This photograph shows the hatch covers in a refrigerated ship. The weather deck covers are of the hydraulically operated end-folding type, arranged in pairs at each end, so that one quarter of the hatchway can be opened. The tween-deck covers are composed partly of two pairs of hydraulically operated covers at each end and partly of two side-folding covers. Various ways of partial opening can be arranged to reduce the heat gain. Two small fork lift trucks can be seen on the gratings of the second tween-deck cover.

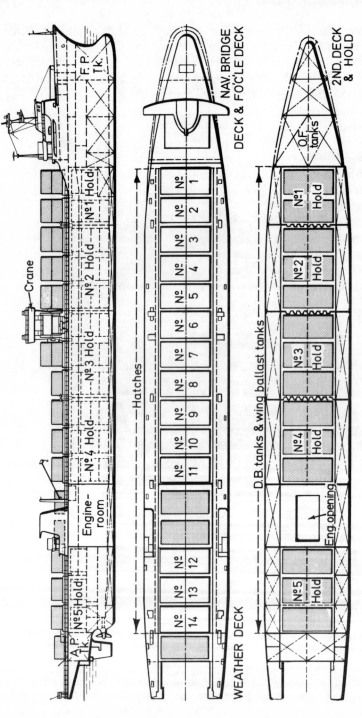

Fig. 8.20. This plan shows the first LASH ship (Lighter Aboard SHip) the 44 000 tdw 19-knot *Acadia Star*. Seventy-three 385-tdw barges can be carried, four-deep in the holds, one- or two-high on deck. The 500-tonne gantry also seen in Fig. 5.54 lifts the barges over the stern and places them in the appropriate holds. Later ships have a slightly greater capacity and speed.

used, with a two-panel pair stowing vertically at each end, and a centre section of two pontoon covers. Fig. 8.19 shows a typical arrangement. One section may be opened to allow the passage of a banana elevator. Insulating material typically consists of polyurethane foam, about 150 mm thick, attached to the underside of the covers. Usually only one tween-deck cover per hold is insulated, dividing the hold into two temperature zones; the other covers are bare steel.

8.4.2 Barge carriers

The two principal types of barge carrier have their barges (lighters) stowed inside their hulls. Others, such as *Bacat*, stow barges on deck or floating between catamaran hulls.

The LASH-type barge carrier (Lighter Aboard SHip), shown in Fig. 8.20, has been the more popular type, with about 30 in service carrying between about 50 and 90 385 tdw barges at 19–23 knots. Deadweight ranges up to about 45 000 tonnes for those ships carrying denser cargoes, although typical cargo payloads in service are more often around 20 000 tonnes, after deducting the weight of the barges. These ships are in essence large cellular container ships, whose 'containers' are barges 18·75 m long × 9.50 m wide × 4·40 m high weighing up to 460 tonnes. Standard containers can also be carried either on deck or in the barges. The barges are stacked athwartships four high in the holds and one or two high on deck, and are handled by a 500-tonne travelling gantry crane which spans the holds and lifts over the stern.

Hatch covers are therefore similar to those fitted on container ships, i.e. pontoon covers measuring about 19·5 m × 10·2 m, one cover per barge stack as illustrated in Fig. 5.54. Corner post stacking points are fitted for the deck-stowed barges. The pontoons are easily handled by the gantry crane and stowed on top of adjacent covers. The hatchways on the barges measure 13·4 m long and 7·9 m wide and are fitted either with end folding pairs, or simple pontoons.

The 'Seabee' type of barge carrier uses larger barges. The first three ships in service carry thirty-eight 850 tdw barges, and two later ones twenty-six 1300 tdw barges. The smaller barges are 29·72 m × 10·67 m × 5·15 m, weighing up to 1020 tonnes fully loaded. They are carried two abreast on three decks above the waterline, and are lifted from the water two at a time by an elevator and moved horizontally on tracks to their stowage location. The upper deck has no hatches, while the second deck is open at its after end to receive the barges. The third (lowest) deck is closed by a watertight door at its after end, as it is only about one metre above the normal waterline.

8.4.3 Heavy lift ships

Although certain multi-deck cargo liners have long been fitted with one or two heavy-lift derricks, it is only since the late 1960s that small vessels have been introduced exclusively for carrying heavy lift cargoes. There are now some 50 ships in the 1000–7000 tdw range with speeds up to 14 knots, either newbuildings or conversions, available for charter to carry indivisible loads weighing up to 1000 tonnes apiece. Such loads include nuclear power plant, petrochemical and oil refining plant, electrical generating components, transport equipment like locomotives and small craft. They are either hoisted on board, from quay or barge, by the ship's derricks (capable of lifting up to 700–800 tonnes), or rolled aboard on heavy lift trailers.

The ship's dimensions are selected to cater for loads which may be over 50 m long, and up to 14 m wide. Arrangements vary between bridge forward, aft or on one side, with derricks forward or on one side. All have as open a deck as possible. A large flush hatch is located in the weather deck so that all but the largest loads can be carried under cover. Its dimensions are generally between 30–50 m long by 9–12 m wide; a similar size hatch is located in the tween-deck if fitted. Cargo may either be stowed on the inner bottom, if its height is below about 6 m so that the weather deck covers can be closed, or if it is too high, either on the weather deck or sometimes on the tween-deck covers with the weather deck covers unshipped.

Pontoon covers, handled by the ship's derricks, span the width of the hatch in about five to twelve panels. They can be made buoyant for floating loads ashore. Design loading varies according to the specification, but the covers may well be able to support distributed loads of up to 10 tonnes/m², or more concentrated locally, e.g. from crawlers. Fig. 8.21 illustrates a modern heavy-lift ship.

8.4.4 Miscellaneous ship types

8.4.4.1 *Passenger vessels*
Side-ports have been the traditional form of access for passengers, crew and stores in all but the smallest passenger vessels. Their design and construction follows the pattern outlined in Section 6.6. But since the opening rarely exceeds 3 m wide by 2 m high, elaborate strengthening and operating mechanisms are not usually required. Where passenger ships also carry cargo, access equipment complies with the same requirements already described for hatch covers and Ro-Ro equipment.

8.4.4.2 *Fishing vessels*
Fishing vessels can be regarded as cargo vessels that load their cargo at sea. They generally have small hinged steel covers on the weather deck to provide access to the fish-rooms, net storage spaces etc. Generally the covers are either flush with the deck or have low coamings, (usually not more than 250 mm) to facilitate the working of nets.

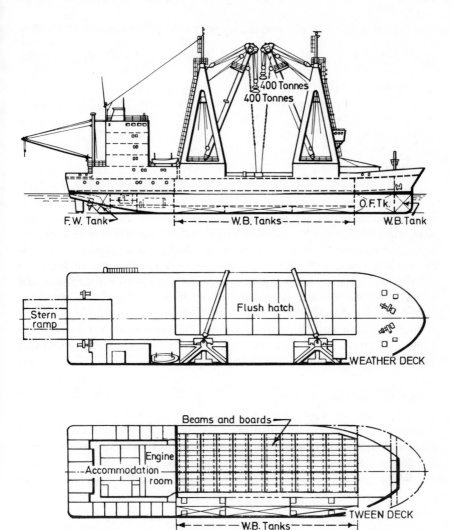

Fig. 8.21. This arrangement of a specialist heavy-lift ship illustrates the clear weather deck with stern ramp and flush hatch. The ship is the 2500 tdw *Gloria Virentium*, fitted with two 400-tonne derricks on the starboard side. The weather deck and the hatch cover are designed for loads up to 6 tonnes per square metre. The seven pontoon sections are handled by the derricks, as are the four sections of the stern ramp, which are stowed on deck when not required for access for rolling loads of up to 1000 tonnes. The tween-deck hatchway is the same size as the weather deck (40·19 m × 11·00 m), but is constructed of sliding steel beams and wooden boards to reduce cost.

A typical modern stern trawler may have the following types of hatches:

(i) Hatch from upper (weather) deck to chute to processing space on main deck for fish, about 3·5 m wide, 1·2 m long – flush steel hinged cover, pneumatically operated.

(ii) Hatch from processing space to insulated fish-room, about 2 m square with insulated slabs in sections on 610 mm coaming.

(iii) Hatch on upper deck above (ii) for unloading in port, about 2·5 m square, with flush steel hinged cover, which will incorporate a trunked section if it passes through accommodation etc.

(iv) Miscellaneous small access hatches to stores, accommodation etc.

8.4.4.3 *Livestock carriers*

Multi-deck ships have long carried occasional consignments of live animals such as cattle, horses and sheep, often on deck, and especially over short routes. Side-ports have sometimes assisted 'walk-on/walk-off' cargoes, but usually no special cargo access arrangements have been fitted beyond the normal hatches. Recently a number of much larger ships have been converted to carry 40 000 or more sheep from Australasia to the Middle East by building light superstructure decks on for example tanker hulls. Light fenced gangways, which pose no special design or construction problems, are used for access onto the ship and between decks.

8.4.4.4 *Offshore supply craft*

The large fleet of supply craft ferrying equipment and stores to offshore drilling rigs and production platforms can be regarded as special types of cargo vessel. In most of them the majority of dry cargo is stowed on the low flat deck aft, where it is accessible by crane from the rig. Cargoes which are stowed below deck are either dry stores or pumpable bulk liquids.

8.4.4.5 *Other types*

Although there is a wide variety of special ship types including icebreakers, cable ships, salvage vessels, tug-barge combinations, training vessels, research ships, etc., their number is comparatively small, and their access equipment is no different in principle from that already described.

Military vessels and naval auxiliaries often have special requirements, e.g. for replenishment at sea, but are outside the scope of this book.

8.5 What sorts of ship are best for general cargo?

8.5.1 General Considerations

The choice of the correct type of ship, its size, speed and equipment, for a particular purpose is not easy. Even when the requirements for the vessel can be clearly defined, e.g. Panamax-sized bulk carrier, there is the problem of choosing the right moment to buy, given expected market conditions. The decisions will vary according to the type of buyer, whether an industrial operator like a paper company needing ships to transport its own products, or an independent owner seeking the most profitable ship to be chartered out on the open market.

For bulk carrying vessels, the choice of type and size often determines the cargo access equipment required, as discussed in Sections 7.1 and 7.2. For general cargo vessels, the choice of ship type is very much wider, while size, speed and equipment need to be selected after consideration of many more factors. As a broad generalization, general cargoes to developed countries consist largely of import substitutes which are cheaper than home-produced goods, while those to developing countries are the only source of manufactured goods.

There is no single answer to the best type of ship for carrying general cargo – it is a question of 'horses for courses', that is, one particular set of circumstances will tend to favour one solution rather than another. A prospective owner will consider a number of alternatives, assess their suitability for his trade from his experience, and compare their economic performance. The following comments are not intended as a substitute for such analyses, but as general guidance to where particular types of general cargo vessel are likely to be advantageous.

The principal factors which influence the choice of ship type, size and speed include the following:

   (i) cargo type and quantity, in both directions;
  (ii) voyage distance and route characteristics, e.g. canals;
 (iii) port facilities, cargo handling methods and costs;
 (iv) infrastructure of countries served, in terms of the national economy and inland transport facilities;
  (v) frequency of service, competition and similar commercial considerations.

Each of the four main types of general cargo vessel – break-bulk, container, Ro-Ro and barge carrier – has a set of circumstances under which it is the ideal choice. Broadly speaking these can be summarized as follows:

   (i) Break-bulk multi-deckers. Suitable for low or irregular cargo flows including semi-unitized cargoes, e.g. pallets, and minor bulks to, or between, developing countries, which have low stevedoring costs, but limited capital available for ships or port investment.
  (ii) Cellular container ships. For balanced flows of manufactured goods on deep sea routes between developed countries, which have high stevedoring costs but good inland transport systems.
 (iii) Ro-Ro ships. For short sea routes for all types of general cargo. For

deep sea routes with a high proportion of wheeled or awkward cargoes, requiring rapid port turnround with limited facilities, especially when congested.

(iv) Barge carriers. For large flows of semi-manufactured medium-value cargoes, e.g. chemicals or steel, on routes serving countries with inland waterway systems, low cost labour and limited port facilities.

Few trades fall exactly into one of these categories, so in most cases it is necessary not only to look at several alternatives, but more particularly to examine hybrid types, e.g. container/Ro-Ro or multi-decker/Ro-Ro, as well as degrees of sophistication within types, e.g. axial versus quarter stern ramp.

Most general cargo vessels operate on liner services, often in Conferences, where a fleet of ships is required to provide a certain capacity in an overall transport chain from producer to consumer, with the port as the interface between sea and land transport. There is a very large number of combinations of ship size, ship speed and number of ships which can provide a particular annual capability. For example, the following fleets of ships, shown in Table 8.2, are all capable of carrying about 75 000 containers a year in each direction on a route about 8000 miles one way.

Table 8.2. Comparison of alternative fleets of equal capacity

No. of ships in fleet	Container capacity, TEUs	Speed, knots	Frequency, days
6	1000	20·5	7
5	1000	25·5	7
5	1250	23·0	8·6
5	1650	15·5	11·2
4	1250	27·0	8·6
4	1650	20·5	11·2
4	2050	16·0	14
3	2050	23·5	14
3	3100	22·0	21
2	3100	28·0	21

In practice, many combinations may be ruled out by physical considerations – for instance draft may be too deep. Commercial considerations are also important; for instance a weekly frequency is desirable on routes like the North Atlantic. Economic considerations such as high fuel bills may rule out very fast ships. Thus the number of feasible fleets to be examined need not be large. It is usually found that ship size is a function of traffic flow, route length, and proportion of sea time to port time; that is, the higher each of these values, the larger the ship. Each end of the spectrum is illustrated by comparing the size of container ships on the Europe–Far East route (over 30 000 tdw), with break-bulk ships on local routes in South-East Asia (under 2000 tdw). The economies of scale are such that big ships offer reduced cost per tonne of cargo on the sea leg of a voyage, but higher costs on the port leg, as time in port increases with ship size unless cargo handling rates can be increased. For decades, this consideration limited the size of break-bulk vessels – increasing their size would only have increased their port time from the existing 50% to an unacceptable 70% or more. Once cargo handling rates about ten times faster became possible with unitization, a major constraint on ship size was removed and, in only a few years, maximum ship size jumped three or four-fold to the current 3000 TEU Panamax ships. Studies [8.3, 8.4] have shown that there is not a great advantage in increasing ship size much further, as it becomes increasingly expensive to provide berths and terminal facilities, and ships exceed the dimensions for transitting the Suez or Panama Canals. Few routes have a traffic flow sufficient to justify container ships much larger than at present, especially when competition encourages a larger number of smaller fleets, e.g. different operators on the North Atlantic each providing a service.

Studies of the economics of speed have shown that most deep sea routes require unit load ships with speeds in the 18–24 knot range, i.e. about 3–5 knots slower than before the quadrupling of oil prices. A further increase in speed is only likely to be justified in special circumstances, as where it becomes possible to provide the required service with one less ship.

Protagonists of a particular type of ship will point to its specific virtues like its cargo handling rate or first cost per unit cargo. While such measures are interesting, they are really only the first step in a detailed evaluation of the alternatives. This must take account of the factors, discussed in Chapter 3, which influence port turnround time under expected long-run conditions.

8.5.2 The cellular container ship versus the Ro-Ro

Over recent years the marine literature has reported the debate on the merits of cellular container and Ro-Ro ships, but there are as yet no final conclusions. On short sea routes, the Ro-Ro vessel predominates because of its fast turnround, flexibility, less expensive terminal requirements and ready door-to-door capability. The small container ship is used either on a few busy routes having a large proportion of industrial goods but a low proportion of wheeled traffic, or as a feeder to deep sea vessels calling at 'mainline' ports.

On deep sea routes with high traffic density, the cellular container ship is superior to the Ro-Ro for carrying a large number of 'boxes', especially on routes where traffic is evenly balanced with fewer problems of empty containers. Expensive berths are necessary, with several gantry cranes, plenty of storage area and terminal equipment like straddle carriers. Such terminals take several years to develop, and investment is lumpy: any extension requires a large additional sum of money over a short period. Very fast handling rates are possible, especially where the ship has a simple itinerary and its length is sufficient for three or four cranes to work simultaneously. With a high proportion of 35 ft or 40 ft containers, up to 2000 containers per day is possible or about 30 000 tonnes of cargo, as in the Sea–Land operation. Under less intensive conditions 400–800 containers per day is more typical say 5000–10 000 tonnes. Thus where the trade has fully accepted the container and stevedoring costs are high, as in the U.S., containers carried in cells are virtually unbeatable, compared with those block-stowed or on trailers in Ro-Ros.

The large Ro-Ro therefore finds its opportunities on routes where there is enough non-containerizable cargo, wheeled vehicles and awkward loads like construction or project cargo to justify a proportion of the total fleet capacity being in this more flexible form. In the ACL solution, this capacity is combined with cellular space in the single hybrid container/Ro-Ro ship (see Section 8.3.3.2), since containers, particularly 40-footers, are such an important element of the North Atlantic trade. On many other routes, not only are containers a smaller proportion of the total traffic, but 20-footers, which are much more easily handled on pure Ro-Ro vessels, predominate.

When traffic flows are lower, the very largest ships are not required. The Ro-Ro suffers from greater dis-economies of scale than the cellular ship, as greater size means more decks, which are more difficult to serve over a stern ramp of reasonable size, with more overstowage problems, and a larger draft variation. The Ro-Ro ship is more expensive per unit of usable volume, i.e. that inside containers or trailers, and its cargo handling rate no higher than cellular ships, rarely more than 8000 tonnes per day. It must therefore compensate by its flexibility – either its ability to serve less developed ports when a quarter or slewing ramp is fitted, or its ability to accept awkward cargoes like heavy lifts on special trailers, which often command a freight premium. In a competitive trade, this versatility is commercially valuable, although in a less competitive one, it may be more sensible to ship the bulk of the traffic in cellular ships, with any appreciable residual traffic going in smaller vessels like hybrid break-bulkers with a Ro-Ro or heavy-lift capability.

An operator with a clearly defined trade can order his ship according to the requirements of that trade, but there is still room for the independent operator who makes tonnage available for charter to provide extra capacity when required. A higher degree of versatility is required for the latter ships, as they may be trading to ports with limited facilities; a good example is the Strider/Tarros class of self-sustaining container ship (see Section 8.2.1). Additional capacity in the longer term can be obtained by building additional vessels, but where frequency of service is an important constraint limiting the number of ships in the fleet, jumboizing may be a better solution.

Ships can be designed initially with jumboization in mind, e.g. lengthening.

In special circumstances, the Ro-Ro ship is often the best choice, at least in the short term. In the U.S. coastal trades to Puerto Rico or Alaska, which are almost entirely road trailer oriented, over-the-road inland deliveries, using trailer ships, are very quick, although there is little opportunity for outside competition. The congestion-beating ability of the Ro-Ro has been well demonstrated in recent years, enabling services to be maintained to ports whose facilities for conventional vessels had been overwhelmed. However, the Ro-Ro is less likely to be the correct long-term solution for the bulk of such trades, once the ports have expanded to cater adequately for other types of vessel, and Ro-Ro capacity no longer commands a freight premium. Road-going trailers and vehicles are valuable when ports are congested, as they can be cleared out of the port area quickly. In more normal circumstances, roll trailers are more suitable where cargo only needs its wheels to move aboard the ship. Such trailers not only have a lower height requirement, but are cheaper despite their heavier payload, since wheels do not need to be to road-going standards, nor are brakes, lights etc. required. Less organization is required by the ship or port operator, since roll trailers do not leave the dock area.

The Ro-Ro ship has gained rapid acceptance in recent years, not only for the advantages already mentioned, but because the ship operator can obtain a greater share of the benefits of an improved service, whereas a comparable improvement in break-bulk shipping requires co-operation and investment by a greater number of organizations who may retain the greater share of any benefits, e.g. if extensive palletization is attempted. The Ro-Ro is also readily switched to alternative routes as demand arises and as developments in a particular trade begin to favour other solutions. Economic studies may show the Ro-Ro to be economically superior under particular circumstances, but the margin is often not great, and it does not need a large change in, for example, cargo mix, for an alternative type to show to an advantage. There will thus continue to be a demand for all the types of Ro-Ro vessel described, including hybrids, but not to the extent that the other types of ship will be completely displaced. Indeed the cellular container ship is likely to remain predominant in many deep sea trades.

8.5.3 Break-bulk cargoes

The break-bulk ship still has a useful role to play in general cargo transport as a comparatively cheap 'work horse', particularly if it has a reasonable container capacity, which increases its attractions for liner operators chartering tonnage for peak periods. It can still challenge unit load ships where efficient cargo handling is practised, especially if palletization and pre-slinging is used (where each cargo package is already prepared in nets or wire slings for rapid handling, and remains so in the hold). However, routes where such efficiency and management skills are likely to be found have already largely gone over to unitization, and shippers have become accustomed to its convenience, its through-transport capability, and reduced damage and pilferage of containers. The break-bulk ship thus finds its role in a number of trades where one or more of the following circumstances apply:

(i) modest or irregular trade flows, or imbalance of cargo in each direction, especially containers with no problems of returning empties, which do not justify the use of specialist or bulk vessels;

(ii) a range of ports to be served, with modest facilities and limited funds for development, often on longer routes with irregular itineraries;

(iii) independence of complex port facilities and restricted inland transport facilities;

(iv) carriage of residual cargoes not taken by container ships;

(v) shipper-packed units readily arranged, e.g. pallets or strapped packages of timber, metal ingots, etc.

(vi) a proportion of bulk or semi-bulk cargoes as well as general;

(vii) plentiful low cost labour, especially where employment is a social factor. Such labour is unlikely to be skilled in the operation of complex equipment;

(viii) cheap ships available, e.g. second-hand, avoiding investment in expensive new specialized tonnage;

(ix) transitional capability required, e.g. semi-container ships can introduce containers to the service in modest quantities, before embarking on full unitization, or multi-deckers having a Ro-Ro capability into the tween-deck.

(x) cargo preference legislation, reserving a share of a trade for national flag carriers who may have limited resources.

The more advanced break-bulk ships with better cargo handling and access equipment, e.g. twin hatches, are capable of working 1000 tonnes of cargo a day in the more efficient ports, which may mean only a day or two in port on a multi-port itinerary where large quantities of cargo are not available. The less advanced ships can serve those less efficient ports where delays to more expensive tonnage would not be acceptable [8.5].

In many cases, cargo carried in the lighters of barge carriers is stowed break-bulk, and many of the conditions described apply. The barge carrier has found it difficult to compete with other unit load ships on routes with a high proportion of containers. Although it is possible to carry containers in lighters, on lighters on deck, in lighter holds or on deck, this operation is slower and more expensive than in cellular container ships. The barge carrier thus has its best prospects in the carriage of fairly large regular consignments of low-value manufactured goods or semi-bulk commodities, like steel, chemicals, forest products, cotton and rice. Where whole barge loads can be delivered to consignees' wharves via inland waterway systems, considerable economies are possible, especially when overland transport alternatives are poor. General cargo to less developed ports can be delivered cheaply since the expensive mother ship is not delayed, while low cost stevedoring labour can load and discharge barges by relatively unsophisticated methods – and overtime working is not necessary.

The number of such suitable trades is however not large, while the mother ships are expensive to acquire for an aspiring developing country lacking investment funds. In many cases, a much better economic return can be obtained by such countries investing their limited funds in improving port and inland transport systems.

References

[8.1] Report on Standardisation of Ro-Ro Ramps. International Association of Ports and Harbours, January 1977.

[8.2] Waage-Nielsen, E. Problems facing Operators of Ro-Ro Systems. *Progress in Cargo Handling*, **6**, Gower Press, 1976.

[8.3] Laing, E. T. *Containers and their Competitors*. Marine Transport Centre, University of Liverpool, 1975.

[8.4] Gilman, S. Optimal Shipping Technologies for Routes to Developing Countries. *J. Transport Economics and Policy*, January 1977.

[8.5] Maritime Transport Research – *Dry Cargo Ship Demand to 1985*. **6**: *Ship Demand*. Graham and Trotman, 1977.

CHAPTER NINE

Specific Design Requirements for Access Equipment

Summary

This chapter describes the specific requirements of the 1966 Load Line Convention where it affects minimum material thicknesses and loadings, as well as giving an outline of the design procedure used for hatch covers and Ro-Ro vessel access equipment. Wheeled vehicles and their axle loadings are described, including the highway regulations for vehicle sizes in major maritime countries. Sealing arrangements are discussed, as well as the materials used in the construction of access equipment. A miscellaneous section at the end of the Chapter describes fire regulations, actuating mechanism power requirements and problems caused by icing.

9.1 Hatch covers

The hatch cover structure consists in essence of steel beams or girders spanning the shorter hatchway dimension, plated over on top and completed by steel side and end plates. The top plate provides the top flange of the beams and girders. For the simpler types of hatch cover panel, e.g. pontoon or single pull, the design and analysis assume a uniformly-loaded simply-supported beam as shown in Fig. 9.1.

9.1.1 Structural regulations

The construction of exposed hatch covers is governed by Regulations 14 to 16 of the 1966 Load Line Convention (1966 LLC). Regulation 14(2) of the 1966 LLC stipulates that exposed coamings and hatchway covers above the superstructure deck shall comply with the requirements of the appropriate national Administration, which means in effect, that such equipment must be designed in accordance with the current practice of the Administration.

Regulation 15(1) specifies the height of weather deck coamings and is discussed in Section 4.5.1. Table 9.1 summarizes the other requirements of these Regulations for beam and board portable covers, and steel pontoon covers secured by tarpaulins and battening devices, and steel covers fitted with direct-securing arrangements. In this context, steel pontoons are plated covers having interior webs and stiffeners, extending the full width of the hatchway and about a quarter of its length, and not fitted with securing arrangements. Classification Society Rules embrace these Regulations but are more detailed and include non-statutory non-exposed cargo covers; they vary slightly between Societies. Some important aspects of hatch cover design are not the concern of the 1966 LLC, such as increased scantlings for special cargo loads.

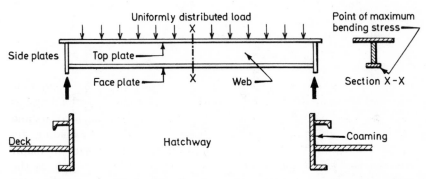

Fig. 9.1. The simplest type of hatch cover structure, where each beam is treated as a uniformly-loaded simply-supported fabricated structure, with maximum bending moment occurring in the middle.

9.1.1.1 *Cleats*

The 1966 LLC requires that satisfactory means for securing weathertightness be provided, and gives requirements for covers secured by tarpaulins and battens. For gasketed covers, details are usually agreed between the hatch cover designer and Classification Society.

An important aspect of the gasket type securing arrangement, whatever its form, is that the pressure between the sealing gaskets and compression bars is correctly maintained. Manual cleats are often over-tightened despite the steel-to-steel contact of the cover skirt plates on the coaming bar, with the result that gasket life is severely shortened. This problem has been largely overcome by the introduction of resilient and quick-acting cleats (Section 5.2.5.2) which means that the correct sealing pressure is consistently and uniformly applied. Cleats on steel covers are generally spaced about 1·5–2 m apart, closer adjacent to the corners, and no more than 0·6 m apart for wooden covers. With the advent of larger and heavier hatch covers, the Classification Societies have, in some cases, approved covers with a greater than usual cleat spacing.

Table 9.1 Summary of 1966 Load Line Convention hatch cover regulations

Hatch cover type	Material	Requirements	Position 1	Position 2
Portable covers secured weathertight by tarpaulins and battening devices	Wood boards and mild steel beams	(1) Minimum thickness of wooden boards – see note (i)	60 mm	60 mm
		(2) Assumed load		
		Ship length under 24 m	1 tonne/m²	0·75 tonne/m²
		Ship length over 100 m	1·75 tonne/m²	1·30 tonne/m²
		Ship length 24–100 m	Interpolate	
		(3) Load factor – see note (ii)	5	
		(4) Maximum deflection	0·0022 × span	0·0022 × span
	Steel pontoons	(5) Assumed load	As for item (2) above	As for item (2) above
		(6) Load factor – see note (ii)	5	
		(7) Maximum deflection	0·0022 × span	0·0022 × span
		(8) Minimum plate thickness	6 mm or 1% of stiffener spacing, whichever is greater	
Other steel hatch covers, secured weathertight with gaskets and clamping devices	Mild steel	(9) Assumed load	As for item (2) above	As for item (2) above
		(10) Load factor – see note (ii)	4·25	
		(11) Maximum deflection	0·0028 × span	0·0028 × span
		(12) Minimum plate thickness	As for item (8) above	As for item (8) above

Notes: (i) With maximum span of 1·5 m. Thickness increased proportionately for larger spans.
     (ii) Maximum stress to be less than (minimum ultimate strength of steel)/(load factor).
    (iii) The strength and stiffness of covers made of materials other than mild steel must be equivalent to those of mild steel to the satisfaction of the Administration.

9.1.2 Loads

The hatch loadings laid down by the 1966 LLC and summarized in Table 9.1 take account of the forces exerted on exposed covers by heavy seas breaking over the deck. Once again however, these are mandatory minimum values and the Classification Societies may require that they are increased in certain cases. Thus for hatch covers on which deck cargoes are rarely if ever carried, the loadings laid down are adequate, typically 1·75 tonnes/m². This is equivalent to cargo stowed 2·45 m high at 1·39 m³/tonne (or 8 feet high at 50 ft³/ton). Loadings for covers in the foremost quarter of the ship's length may be increased to counter the severe sea water forces likely to be experienced in this region.

9.1.2.1 *Deck cargo*

Where containers or additional deck cargoes are to be carried on exposed hatch covers, the actual loads expected must be used for design purposes if they exceed the basic requirement. It is usual to assume uniformly distributed loading (although in the case of containers, point loading, see below) so the calculations must take into account the stresses imposed in the cover plating between transverse beams as well as in the beams themselves. Deck cargo is, however, often stowed on the decks alongside the hatchways; thus, with the exception of timber and containers, stowage of cargo on hatch covers, especially peak-topped, is normally avoided. The surrounding deck structure must also be designed to support any additional loads.

Tween-deck and flush weather deck hatch covers are usually designed to take fork lift truck wheel loads, as it is becoming increasingly common to handle general cargo in this way. Generally, small fork lift trucks are electric or have propane-fuelled internal combustion engines, and weigh in the region of 3–4 tonnes unloaded. Thus it is common practice to design tween-deck covers to withstand an axle load of 6–8 tonnes, since most of the weight of a fully laden truck is taken on the front axle. Axle loads for various payloads are discussed further in Sections 6.1.2 and 9.2.2.

9.1.2.2 *Containers*

Hatch covers on which containers are stowed must be designed to withstand point loads transmitted through the container corner fittings, if these are more severe than the uniformly distributed load specified in the Rules. The entire weight of a container (and those stacked upon it) is transmitted through the corner posts. It is usual for the transverse beams of the cover to be located directly below the corner fittings in order to give the maximum support. Hatch side fittings may require special stiffening. The dynamic effects of heave, pitch and roll are also allowed for. The cover top plating is usually stiffened by shear-resisting pads under the corner fittings, which are often raised on stools to avoid piercing the top plate and to ensure that they are self-draining.

Only those corner fittings that are positioned away from the hatchway coaming require substantial support, as those near the periphery of the hatch cover transmit their loads directly to the coaming through the side and end plating. Thus it is possible that a 41 ft (12·5 m) long hatch cover loaded with 40 ft containers requires no appreciable extra stiffening and is virtually the same weight as a normal cover. 20 ft containers would require intermediate support, however; Fig. 7·5 gives some guidance on the extra weight involved.

9.1.2.3 *Liquid cargoes*

Combination carriers are built with large, heavy oiltight covers which, besides keeping the sea out, must keep the cargo in. Both hydrostatic and hydrodynamic forces may be exerted on the interior surface of the cover.

1. Hydrostatic forces. When a ship is rolling and pitching, liquid cargo may press intermittently against the internal surfaces of the hatch covers. In a 14 m-wide hatchway, a roll angle of 13° may result in a pressure on the side of the cover of 1·7 tonnes/m² (Fig. 9.2) [9.1]. In most cases, however, with only one hatchway per hold, the free surface bears on the underside of the deck plating and not on the hatch cover when the ship is pitching, but with two or more hatchways (as in ore/oil carriers) pitching becomes more important. An additional factor is the presence of internal gas pressure from the protecting inert gas blanket which, although controlled by a pressure/vacuum relief valve, can reach the set pressure of 1·4 tonnes/m² [9.1]. When

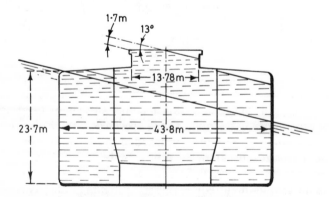

Fig. 9.2. Assumed rolling in an ore/oil carrier can induce appreciable hydrostatic pressures on the underside of oiltight hatch covers.

the holds are completely empty after carrying an oil cargo, they are also usually filled with an inert gas generated from an on-board system which is designed to run at a pressure of about 1·5 tonnes/m².

2. Hydrodynamic forces. These forces are caused by liquid cargo moving within the hold and sloshing against the hatch covers. Bureau Veritas has calculated [9.1] that pressures of 6·7 tonnes/m² can be reached in extreme conditions, in a typical large combination carrier.

It is not usually necessary for the designer to calculate these forces for each ship as they are taken into account by the Classification Society Rules for the determination of hatch cover and cleat scantlings.

9.1.3 Scantlings

Minimum scantlings (thicknesses and dimensions of steel plating and stiffeners) and load factors (or factor of safety) for steel covers are specified in the 1966 LLC, so that the load factor multiplied by the maximum stress in the cover is less than or equal to the minimum ultimate strength of the steel. Corresponding minimum scantlings depending on loading are given in Classification Societies' Rules; alternatively direct calculations of required structural strength may be made using maximum design stress levels.

Table 9.2 shows the requirements of Lloyd's Register with respect to steel hatch covers or other access equipment constructed of Grade A mild steel having a minimum ultimate tensile strength of 4 100 kgf cm² [4.6].

Table 9.2. Access equipment strength calculations

Factor	Weather deck	Tween-deck
Maximum bending stress, kgf/cm²	965	1200
Maximum shear stress, kgf/cm²	700	700
Maximum deflection	0·0028 × span	0·0035 × span

The top plate of a typical hatch cover panel may be from 6–13 mm thick depending on the spacing of the beams (generally, thickness = spacing/100). This plate is stiffened by beams spanning the hatch cover, usually fabricated 'tee' beams having a depth of about 4% of the span and spaced 500–1000 mm apart. The panel is completed by side and end plates which may be from 8–20 mm thick.

Double-skinned panels are sometimes employed. These are of two basic types. The first uses a thin bottom plate about 4–5 mm thick and usually un-stressed, while the second has a bottom plate of equal thickness to the top plate which shares the stresses in the cover. It is claimed that double-skinned panels are easier to paint, clean and maintain than single skin panels. Such covers can also reduce the extent to which cargoes like grain can shift within the hatchways. The void space between the skins must be treated to reduce internal corrosion.

A disadvantage of double-skinned covers is that on large hatchways they may be over-stiff. This can lead to sealing problems when the ship is working in a heavy sea, and the hatch coamings are deflecting; this is discussed further in Section 9.1.4.

Double-skinned panels are more expensive to manufacture than single skin. They are usually constructed by laying the top plate upside down and welding the stiffeners to it. These are followed by plates welded to the stiffener face bars (Fig. 9.3a). Alternatively, the face bars may be dispensed with as illustrated in Figs. 9.3b and 9.3d. On large covers, manholes may be cut in the bottom plate between the webs to allow the welder entry into the space between the skins as shown in Fig. 9.3c.

An alternative to double skinning where ease of cleaning is necessary, is to fit round section face bars to the webs thus avoiding the 'shelf' formed by a rectangular section bar. This is shown in Fig. 5.40.

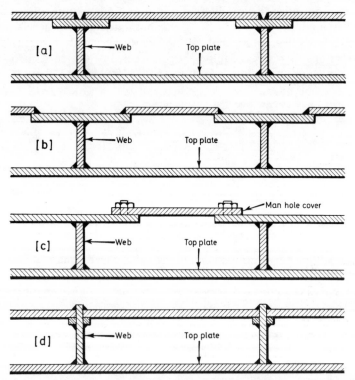

Fig. 9.3. Alternative methods of constructing double-skinned hatch covers. Covers are shown upside down, as normally fabricated.

9.1.4 Deformation

As ships become larger and hatchways take up a greater percentage of the deck area, so the question of hatchway deformation becomes more important. Traditionally ships have had fairly small hatchways and so have derived a considerable amount of their strength from their decks. As hatchways have increased in width, so the deck's contribution to the longitudinal and torsional strength of the hull girder has declined, being limited to the strips of deck outboard of the coamings and between the hatches. A ship with hatchways more than 70% of the beam in width has approximately half the torsional rigidity of a similar ship with hatchways which are only 40% of the ship's beam. Compensation, in the form of thickened plating and/or box girders may be required.

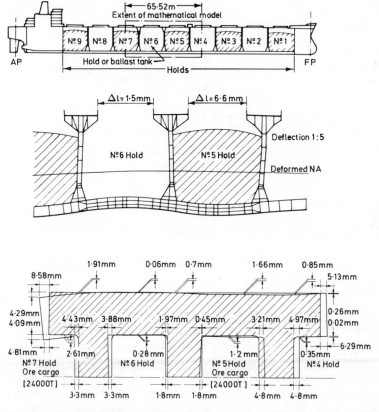

Fig. 9.4. Detailed analysis of the structure of a 120 000 tdw bulk carrier by Lloyd's Register indicates the extent of deformations resulting from the loading of ore in alternate holds in still water conditions. No. 5 hatchway increases in length by 6·6 mm, but reduces in width by 2·4 mm.

Fig. 9.4 [9.2] indicates the extent of hatchway deformation that may be encountered in a large bulk carrier loaded in alternate holds. Although the deformations are not excessively large, and are not permanent, they can be sufficient to allow sea water to enter, and parts of the hatch covers to fracture.

It has been suggested that, since hatchways are always closed by hatch covers, the latter ought to be designed to contribute to the strength of the hull girder. This is, however, not practical as the devices securing the cover to the coaming would have to be excessively robust, and any distortion due to the loading of the vessel in port could lead to the covers becoming jammed and thus impossible to open. Moreover, to make a significant contribution to a vessel's strength, the covers would have to be considerably stronger and heavier than at present, and this could introduce operational difficulties.

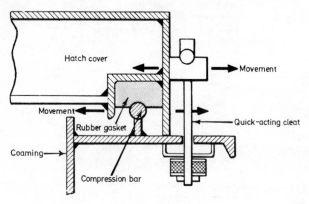

Fig. 9.5. Gaskets and cleats must be able to accommodate the longitudinal (and transverse) distortion of hatch cover panels.

Hatch coamings can be designed to contribute to the longitudinal strength of the ship but there can be problems associated with this. If it is decided to use the coamings in this way, they must be continuous over the midship portion of the ship and properly tapered at the forward and after sections of the hull. The longitudinal material of the coaming is then taken into account in calculating the midship section modulus, according to Classification Society requirements. But since the maximum distance from the neutral axis is also increased, there may be little net benefit. If however the coamings are not intended to contribute to the hull strength, positive steps have to be taken to ensure that the adverse effects of discontinuities are minimized. Side coamings are often constructed as continuous girders on large ships with wide hatches, e.g. container ships or forest product ships, but in most other ship types, coamings are made discontinuous.

Hatchway coaming deflections in service are taken into account in hatch cover design in the following ways.

9.1.4.1 *Longitudinal deformation*
Longitudinal deformation of the top of the coaming is due to the hogging or sagging of the vessel. It depends on hatchway length and may be as much as 7–8 mm at each end.

Fig. 9.6. Frictional forces resulting from steel-to-steel contact between pontoon covers and coaming bars on hatch covers can result in fractures to the side plates on container ship hatches with their heavy deck loads, unless bearing pads are introduced.

Longitudinal deformation is compensated for by fitting the ends of the hatch cover with wide gaskets whose rubber absorbs the relative movement of the compression bars as the ship works. Hatch-end cleats must allow such movements, which would otherwise be taken up at the cross-joints with attendant risks of leakage (see Fig. 9.5). This illustration also shows the steel-to-steel contact between the hatch cover and coaming. The purpose of this is to prevent the over-compression of the gaskets. However, it gives rise to a frictional force whose magnitude depends on the pressure of the cover on the coaming bar and their relative movement. Fig. 9.6 shows the effects of this force on the one-piece cover of an early design of container ship. This problem has since been remedied by the addition of mild steel bearing pads between the hatch cover main beams and the coaming, thus ensuring that the cover side plates are not supporting the weight.

9.1.4.2 *Transverse deformation*

Transverse deformation is caused mainly by a vessel's changes in draft as cargo is worked, (see Fig. 9.7) but also by hogging or sagging. Deformations of as much as 15–25 mm over the width of long hatchways are possible. Relative transverse movement between the cover and coaming is allowed for in a similar way to longitudinal deformation, by providing the cover with wide gaskets, and by ensuring that the wheels on one side of the hatch cover panels are free to move laterally, as illustrated in Fig. 9.8.

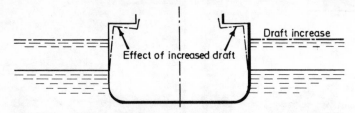

Fig. 9.7. The increase of hydrostatic pressure from deeper drafts can cause transverse deformations of hatchways.

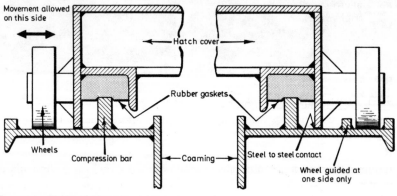

Fig. 9.8. Transverse deformation can be accommodated by restraining wheels on one side of the cover but not on the other.

9.1.4.3 *Torsional deformation*

Torsional deformation is caused by a combination of non-symmetrical loading of the vessel and hydrodynamic forces when the ship heads diagonally into waves. It can rack the midships hatchway diagonal of a large container ship by as much as 30 mm [9.3]. Fig. 9.9 illustrates torsional distortion and the resultant movement between the hatch cover and coaming. By fitting resilient cleats at the ends of the hatch covers, some movement between cover and coaming takes place, thus reducing the relative movement between panels. Typically the movement at the cross-joint, a, is up to 25 mm and that at the ends, b, in the region of 15 mm, on a large bulk carrier. On a hatch fitted with a single-piece cover, or on two-panel covers rigidly connected when in the closed position, all movement must take place at the ends of the hatchway. Once again relative movement is allowed for by the provision of wide gaskets.

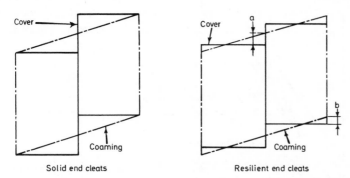

Solid end cleats   Resilient end cleats

Fig. 9.9. Torsional distortion of hatchways causes greater relative movement at cross-joints with solid end cleats (left view) than at those with resilient cleats (right view).

9.2 Access equipment in horizontal loading ships

9.2.1 Structural regulations

At the present time Ro-Ro vessels are regarded as ferries or cargo ships by the Classification Societies. The 1966 LLC treats bow, stern and side doors in Ro-Ro vessels as access openings into an enclosed superstructure and requires them to be framed, fitted and stiffened so that the whole structure is of equivalent strength to the unpierced shell or bulkhead, and watertight when closed. Openings in the freeboard deck have already been discussed in Section 4.3.2.

The Classification Societies require the cleating arrangements to be such that the connection between the door and the surrounding structure is of sufficient strength and publish detailed requirements. Cleats on bow and stern doors should be no more than 1 m apart if hydraulically operated, or 0·6 m if manually operated. Spacing for openings in vehicle decks should not exceed 1·5 m.

9.2.2. Loads

The loads assumed to act on bow visors, bow and stern doors, side ports and bulkhead doors are those acting on the surrounding structure. Scantlings for plating and framing are laid down in Classification Society rules in the same way as those for shell or bulkhead structure.

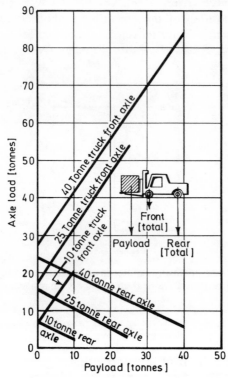

Fig. 9.10. Fork lift trucks, especially those carrying containers, impose high axle loads on decks and ramps. Most of the load, which increases directly with increasing payload on the vehicle, is taken on the front axle.

Ramps, hatch cover/ramps and elevators are traversed by wheeled vehicles and are subject to overall gross loads and point loads applied by individual wheels. General characteristics of typical vehicles in Ro-Ro ships are shown in Table 6.1. Figs. 9.10 and 9.11 show how axle loadings vary with payloads less than the maximum for two common types of wheeled equipment: fork lift trucks and roll trailers. These curves give total axle loads; wheel loads can be estimated by dividing the total by the number of wheels on an axle of a particular item of equipment. If, as is common with some trailers, there are two or more axles very close to each other at the rear, they are assumed to share the load equally. Table 9.3 indicates the tare (empty) weight, maximum cargo payload and maximum gross weight for various standard container types which can be carried on such equipment. Table 9.4 gives the maximum dimensions and weights of road-going vehicles in major maritime countries. The limits shown in Table 9.4 may restrict the maximum payload of containers (especially 40 ft containers) carried on the highways of certain countries.

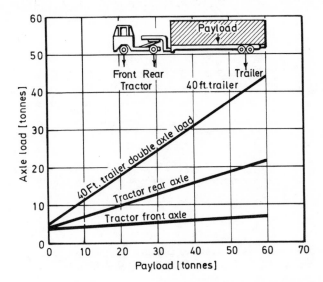

Fig. 9.11. Roll trailers, especially 40 ft (12 m) types, also impose high axle and wheel loads, because of their small diameter, closely spaced, solid rubber tyres.

These payloads can be used in conjunction with the axle load curves given Figs. 9.10 and 9.11. For example a 20 ft steel container, loaded with 18 tonnes of cargo weighs about 20 tonnes and when handled by a 25-tonne capacity fork lift truck gives rise to a front axle load of about 47 tonnes and a rear axle load of about 5 tonnes. The precise figures depend upon the model of truck being used and the weight of the spreader attachment, but it is obvious why many Ro-Ro ships are designed for 50–60 tonne axle loads.

When a trailer or fork lift truck is in motion, dynamic effects are assumed to be about 30–40% of the static load [9.4]. Thus ramps and deck plating are designed to withstand a load of 1·4 times nominal load. The manufacturer and shipbuilder will of course take such considerations into account when designing the structure of access equipment, decks and supports to suit the specified loads.

9.2.3 Scantlings

Classification Societies vary in their requirements for the thickness of deck plating to support wheeled vehicles. Besides the use of standard formulae

Table 9.3 Typical container characteristics.
All containers 8 ft wide (2·44 m)

Type	Length, ft (m)	Height, ft-in (m)	Internal capacity, m³	Tare weight, tonnes	Max. cargo payload, tonnes	Max. gross weight, tonnes
Dry freight, aluminium	40 (12·19)	8–0 (2·44)	63·3	2·80	27·68	30·48
Dry freight, aluminium	40 (12·19)	8–6 (2·59)	67·0	3·40	27·08	30·48
Dry freight, aluminium	40 (12·19)	9–6 (2·89)	75·0	3·90	26·58	30·48
Dry freight, steel	40 (12·19)	8–0 (2·44)	63·0	3·40	27·08	30·48
Dry freight, steel	40 (12·19)	8–6 (2·59)	67·0	3·60	26·88	30·48
Dry freight, aluminium	35 (10·67)	8–6½ (2·60)	59·2	2·50	22·67	25·17
Dry freight, steel	30 (9·12)	8–0 (2·44)	46·0	3·00	22·40	25·40
Dry freight, aluminium	20 (6·06)	8–6 (2·59)	33·0	1·90	18·42	20·32
Dry freight, steel	20 (6·06)	8–6 (2·59)	33·0	2·20	18·12	20·32
Dry freight, steel	20 (6·06)	8–0 (2·44)	31·0	2·00	18·32	20·32
Dry freight, steel	10 (2·99)	8–0 (2·44)	14·7	1·30	8·86	10·16
Open top, steel	40 (12·19)	8–6 (2·59)	65·0	4·30	26·18	30·48
Open top, steel	40 (12·19)	4–3 (1·30)	27·0	3·90	26·58	30·48
Open top, steel	20 (6·06)	8–0 (2·44)	29·3	2·10	18·22	20·32
Insulated	40 (12·19)	8–6 (2·59)	61·0	4·50	25·98	30·48
Insulated	20 (6·06)	8–0 (2·44)	27·0	2·30	18·02	20·32
Refrigerated*	40 (12·19)	8–6 (2·59)	56·0	5·80	24·68	30·48
Refrigerated*	20 (6·06)	8–0 (2·44)	24·0	3·30	17·02	20·32
Tank	40 (12·19)	4–3 (1·30)	21·4	4·20	18·48	22·68
Tank	20 (6·06)	8–0 (2·44)	19·1	2·80	20·85	23·65

*Integral refrigerating unit.
Note: Nominal dimensions in feet, actual in metres.

Table 9.4 Guide to road-going vehicle characteristics in selected maritime countries

Characteristic	E.E.C.(1)	U.K.	Italy	Greece Turkey(a)	Poland Yugoslavia Portugal(b)	Sweden	U.S.S.R.	U.S.A.(2)	Japan	Australia(3)
Maximum height, m	4·0	NS	4·0	3·8	4·0	NS	3·8	4·11	3·8	4·3
Maximum width, m	2·5	2·5	2·5	2·5	2·5	2·5	2·5	2·44	2·5	2·5
Maximum length, m										
Rigid lorry	11	11	11	11	12	ca 12	12	NS	12	11
Articulated lorry	15	15	14	14	15	ca 18	20	16·8	NS	15·3
Lorry with trailer	18	18	18	18	18	24	24	19·8	25	16·8
Max. axle load, tonnes										
Single	10	9·1	10	8, 10(a)	10	10	10	9·1	10	8·4
Double	16	18·2	14·5	14·5	16	16	20	16·3	20	14·0
Max. vehicle weight, tonnes										
Rigid lorry	22	24·4	18	19, 16(a)	24, 22(b)	ca 30	25	26·7	20	23·4
Articulated lorry	38	32·5	32	36	38	ca 40	40	33·1	NS	33·9
Lorry with trailer	38	32·5	32	38	38	51·4	33	33·1	NS	37·5

NS = Not specified.
N.B. Detailed regulations are usually complex, with precise values depending on axle configuration etc. The figures are given as an indication of typical values.
(1) Typical values for E.E.C. countries except those in other columns.
(2) Varies between states; typical values for maritime states shown.
(3) Varies between states; New South Wales given.

they will also accept scantlings determined as a result of the direct calculation of stresses, provided that they are submitted for inspection together with such data and assumptions as are used in them. The limits given in Table 9.2 also apply to Ro-Ro access equipment.

Some Societies, such as Lloyd's Register, take into account the tyre print dimensions whereas others are only concerned with whether the vehicle has pneumatic, solid rubber or steel tyres. Table 9.5 gives approximate plate thicknesses for various fork lift truck payloads, and is intended as an early guide to plate thickness when insufficient data regarding axle loading and wheel dimensions are available. When designing suitable deck plating, the relevant Classification Society rules must be adhered to; each may differ in detail.

Table 9.5. Approximate deck thicknesses for fork lift trucks

Fork lift truck payload, tonnes	Maximum s/t s = actual frame spacing t = plate thickness	Plate thickness to nearest 0·5 mm for stiffener spacing of:	
		s = 600 mm	s = 900 mm
1	85	7·0	10·5
3	55	11·0	16·5
5	45	13·5	20·0
10	37	16·0	24·5
20	32	19·0	28·0

It is important for an operator to specify the full range of wheeled equipment that he expects to use on board his ship, so that access equipment and decks are made strong enough for all anticipated services. Different types of equipment may determine the scantlings of the various parts of the structure, for example, wheel loads and tyre prints influence plating thickness, while axle loads and spacing influence beam and girder scantlings, and gross vehicle weights influence overall strength and operating mechanisms. However far-sighted a ship operator is, when his ship is in service, he will inevitably be asked to transport cargoes with load characteristics not covered in the original specification. Once a piece of access equipment, e.g. an axial stern ramp, has been designed, it is possible to calculate the full range of axle loadings that it can withstand. The operator can then use a diagram such as that shown in Fig. 9.12 to assess whether the proposed load is within the capability of the equipment.

9.2.4 Hull deformation

Hull deformations are not usually a serious problem in Ro-Ro ships so far

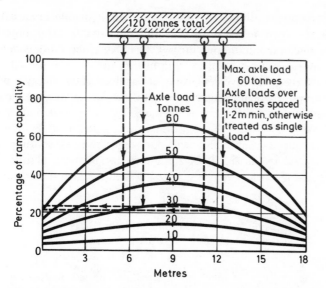

Fig. 9.12. A ramp is designed to withstand certain specified loads, e.g. axle loads. Once built, it will have a capability for a wider variety of other load configurations. Proposed axle loads and spacings can be checked on a diagram such as that illustrated, to determine whether the total is within the ramp's capability, i.e. below 100%.

The ramp is a hypothetical design of axial stern ramp 18 m long, constructed of multiple parallel longitudinal girders. The example shows a vehicle with a total weight of 120 tonnes distributed over four axles. When the vehicle is at the middle of the ramp the total load is about 90% of the maximum. Individual wheel loads and tyre prints would also need to be checked against plating panel strength.

A slightly more detailed assessment would be required for quarter ramps with deep side girders.

as access equipment is concerned. Most Ro-Ro vessels do not have large openings in their weather decks and hence are of a closed box type of construction which is adequately stiff. When openings are cut in the side shell for doors, adequate compensation is required within the midship half length. Not so much compensation is required in the less highly stressed sections of the hull, i.e. the forward and after quarters of the overall length.

Bulkhead doors and large stern doors must be arranged so that they do not become jammed by the racking of their frames. The design of transverse bulkheads, and the operating mechanisms of such doors, take this into account.

9.3 Advanced design techniques

Design invariably involves compromises between conflicting requirements.

All shipboard structures must be strong enough to satisfy regulatory requirements, but the designer also has to take into account a wide range of
operational, technical and economic factors. For the structural design, the
designer has two alternatives open to him: he can use the traditional and
well-tried manual methods of calculation, often distilled into simple procedures, or he can use more advanced techniques, usually requiring the assistance of computers.

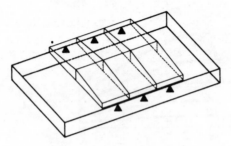

Fig. 9.13. The structure of hatch cover panels supported on two sides is
straightforward to analyse.

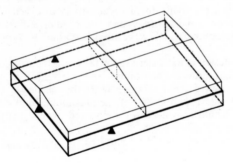

Fig. 9.14. Structure supported on three sides is more complex to analyse with
transverse and longitudinal members.

Fig. 9.13 illustrates a single pull type hatch cover panel, supported along
both sides and transversely framed, whose structure is easy to analyse by
hand as discussed in Section 9.1. No great advantage would be gained by
using a sophisticated computer-based design procedure to minimize the
weight of such a structure. A more complex arrangement is illustrated in
Fig. 9.14 where the cover shown is supported along three edges and is framed
both longitudinally and transversely. Such a grillage could be analysed
manually but it would be a lengthy process and hence is more suited to
computer analysis. In complex structures deflection under load is important,
for example, to ensure that the torsional stiffness of a quarter ramp is such
that it bears evenly on the quay despite heeling of the ship. It is now
normal practice to use a finite element analysis program in such circum-

stances. This splits the structure up into small elements as shown in Fig. 9.15 and calculates the stress and deflection in each element and its effect on the structure as a whole. This gives the designer a detailed view of the structure's performance in terms of stresses and deflections of a proposed arrangement.

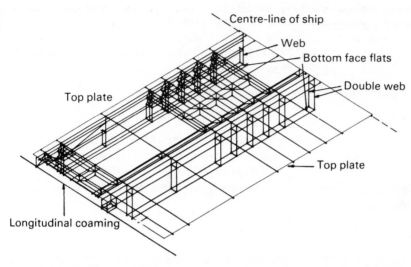

Fig. 9.15. Grillage structures composed of multiple beams and girders with support on several sides and with varied loading patterns are best analysed by computer methods, including finite element techniques.

Computer programs may include optimization procedures, so that the best combination of design variables – dimensions, spacing of stiffners, scantlings of elements – is automatically calculated. Once the loading pattern and dimensions of the hatch cover or ramp have been specified, such programs search through all their feasible combinations to optimize the structure with respect to, say, minimum weight, subject to certain specified constraints such as panel size, deflection etc. This can be done by supplying the program with an outline design which it analyses in terms of stress levels. It then adjusts the scantlings so that no element is overstressed. The final procedure of the program can be to increase the new scantlings it has found to standard sizes of beams and plates.

Computers can be used in several different forms, ranging from desk-top computers with stored programs for simple problems, through central computers linked to the designer's office via the telephone network, to the most elaborate systems, including graphics units so that the designer can control, display or modify the data and results. The potential advantages of using such advanced design techniques are given below:

(i) more factors can be taken into account, with fewer approximations;

(ii) a wide variety of alternative configurations can be investigated;

(iii) time can be saved during the design stage with information produced more quickly;

(iv) technical factors can be combined with economic factors to produce an overall optimum design;

(v) the program can be stored and re-run at any time. For example if the operator of a Ro-Ro ship is offered a cargo with an unusual axle configuration, the ramp manufacturer can quickly advise whether the load is within the designed capability.

The end result is a better and more efficient product.

9.4 Seals

The 1966 LLC requires cargo access openings to be weathertight, as discussed in Section 4.1.5. Chapters 5 and 6 indicate that weathertightness is usually achieved by an arrangement of sealing gaskets and drainage channels, and illustrations of typical installations are given there. The main requirements of a sealing system are as follows:

(i) it must prevent any transfer of water from outside the ship to the cargo space;

(ii) on combination carriers, it must be oiltight to prevent any transfer of liquid from the inside of a cargo space to the outside, when subjected to the pressures outlined in Section 9.1.2.3;

(iii) it must be able to maintain the weathertight integrity of the cargo

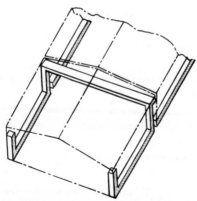

Fig. 9.16. Arrangement of the rubber gasket around the edges of two adjacent single pull panels.

space in all sea states and thus, it must be able to accommodate the deformations discussed in Section 9.1.4;

(iv) it should be resilient and able to accommodate normal irregularities in the mating surfaces;

(v) the gasket should be abrasion resistant as it may rub against the mating surface. It must also be resistant to cargo contact, e.g. oil on combination carriers;

(vi) it should be easy to maintain;

(vii) it should retain all the above properties throughout a long service life, exposed to climatic extremes.

Fig. 9.16 shows the usual arrangement of the seal round the edges of two adjacent panels of a single pull type hatch cover.

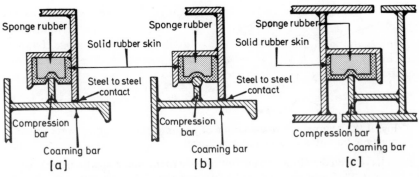

Fig. 9.17. Conventional gasket and compression bar arrangements; (*a*) peripheral gasket with rectangular bar; (*b*) peripheral gasket with round bar; (*c*) cross-joint gasket.

9.4.1 Gasket arrangements for hatch covers

The most common form of sealing arrangement employs a compression bar which bears against the 'rubber' gasket as illustrated in Fig. 9.17; the actual material may not always be rubber.

The gasket's skin is abrasion resistant and protects the underlying softer, resilient section, e.g. sponge rubber, which allows the gasket to deform sufficiently to effect an efficient seal. The hatch cover side plates rest against the steel coaming so that only such weight is borne by the gasket as is needed to achieve tightness. This ensures that the gasket is not over-compressed – a condition that would result in a permanent set in the rubber. It has been found that a permanent compression set will not usually result if the designed area after compression is not less than 75% of the uncompressed cross sectional area of the rubber. Other arrangements use slightly different gaskets, some of which are illustrated in Fig. 9.18. In all these the rubber is bedded into a special marine type of adhesive in the gasket retaining channel.

Although the gaskets illustrated in Fig. 9.18 are elastic in the vertical plane, they are very stiff in the horizontal plane. In Section 9.1.4 it was noted that horizontal displacements may be substantial. Consequently special

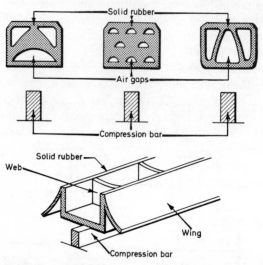

Fig. 9.18. Alternative cross-sections of gasket types permitting vertical compression, but little horizontal displacement.

gaskets have been developed to accommodate them, some of which are shown in Fig. 9.19. Those shown in Figs. 9.19a and b are airtight. When the hatch is closed the gasket is compressed and its internal air pressure rises from about 1 bar (1·02 kg/cm²) to 1·8 bar. A variant of this is a gasket with an external air supply by which the internal pressure is raised to about 1·8 bar, when the hatch is closed.

The gaskets illustrated in Figs. 9.19c and d have no compression bars, but

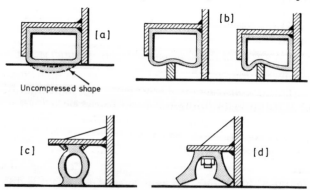

Fig. 9.19. Gasket types permitting horizontal as well as vertical displacements.

simply rest on the flat top plate of the coaming. They form a relatively inexpensive seal because the rubber section is easily extruded and requires no gasket channel. These last two arrangements permit considerable horizontal displacement of the hatch cover, but the absence of compression bars demands that they be maintained in good condition at all times.

A unique solution to the problem of sealing combination carrier hatch covers, which allows the relative movement between hatch cover and coaming to be taken by a permanently fixed rubber skirt, has been developed. The seal between the skirt and the coaming is made by a flexible steel frame which follows the movement of the coaming as the ship works. The weight of the cover is taken partly by the seal but mainly through the cleating 'grippers' described in Section 5.6.6. This arrangement is shown in Fig. 5.41. A similar, permanently fixed seal has already been discussed in Section 5.5.6 on Rolltite cover seals. There are also devices which combine sealing and cleating and two of these are illustrated in Fig. 9.20.

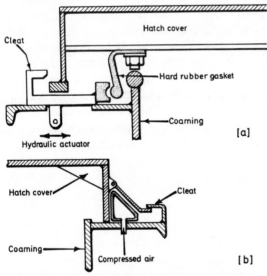

Fig. 9.20. Two arrangements which combine sealing as well as cleating.

It is important that compression bars and gasket channels, where used, are accurately aligned during construction. The amount of compression that a gasket can be subjected to before suffering a permanent set is limited and, in practice, only very small deviations from a straight and level line can be tolerated for the contacting surfaces. Manufacturers check that the tolerances, as installed by the shipyard, are within acceptable limits.

9.4.2 Gasket materials

Gasket material must be of a suitable quality; it should not harden excessively when subjected to sub-zero temperatures nor soften in tropical conditions. Particularly important in this respect are the gaskets fitted to refrigerated vessels.

Most gaskets used for seals on dry cargo ships are of natural, synthetic or neoprene rubber, while combination carriers require a nitrile composition, resistant to chemical attack by oil. In general neoprene synthetics have good heat, ageing, weather and flame resistance, but only moderate oil and chemical resistant properties. Nitrile rubbers have better oil and chemical resistance but worse cold temperature properties.

The general construction of seal steelwork, welding and painting needs careful attention.

9.4.3 Compression bars

The usual arrangement of the compression bar is shown in Fig. 9.17a; it is of rectangular section mild steel welded to the coaming bar. After several years of service, these bars can become badly corroded and require replacing. Moreover, they may then have sharp corners which press into the gasket so that the rubber takes on a permanent set earlier in its life than would be the case for a sealing arrangement employing either a round compression bar, as shown in Fig. 9.17b or a rectangular one with rounded corners. The latter produces an element of 'knife-edge' loading which gives a good seal with the rubber. When such compression bars are fitted, they are often of stainless steel, which, although initially expensive, is often worthwhile as it is not usually necessary to replace the bar during the life of the covers. Section 7.3.2 shows how the benefits of such a fitting can be evaluated economically. Zinc spraying of mild steel compression bars can also be used for the same purpose.

9.4.4 Gasket arrangements for Ro-Ro access equipment

Doors, ramps and bow visors must be equipped with weathertight sealing arrangements. These employ similar gaskets to those of hatch covers described in Section 9.4.1. Since the deformation of the openings to which they are fitted is small (as discussed in Section 9.2.4) the simpler types of seal are usually adequate.

9.5 Construction materials

In general, Grade A mild steel is used for the construction of all cargo access equipment. Grade A is the 'ordinary' mild steel used for most shipbuilding purposes. Certain applications require Grades D and E steel. Grade D is a notch-tough steel with a chemical composition largely chosen by the steel producer provided that it retains good weldability. Grade E steel is the highest grade and is also notch-tough but its manufacture and composition are strictly controlled. Grades D and E steel are specified for coaming bars on refrigerated ships where low temperature brittle fracture must be guarded against. As these steels have low corrosion resistance, special coatings, e.g. epoxies, are often used on the underside of hatch covers where sweating can cause accelerated corrosion.

Other materials such as higher tensile steel and aluminium are acceptable to the Classification Societies.

9.5.1 Higher tensile steel (HTS)

If the stress level within a structure were the only factor governing its scantlings, the use of higher tensile steel instead of mild steel would result in weight savings of up to 15%. However stress is not the only governing factor. Deflections, minimum thicknesses, ease of construction, initial and maintenance costs must all be taken into account. Higher tensile steel (HTS) can sometimes be used for the beams of hatch covers, with their top plates made from mild steel. This reduces the weight of the covers, without incurring the extra cost of using all HTS construction. Thus, HTS is sometimes used on hatch covers where, by reducing the cover weight slightly, it may be possible to install less powerful operating mechanisms. Construction entirely in HTS is sometimes used in pontoon covers for container ships, to keep their weight below the maximum lifting capacity of container cranes.

9.5.2 Aluminium

Aluminium structures can be 55–60% lighter than equivalent mild steel structures. Most of the problems associated with higher tensile steel are not present with aluminium. Since it is not as strong as steel, thicknesses have to be increased, thereby alleviating problems associated with minimum regulatory deflections, buckling etc., unless the design is such that deflection is significant. Welding aluminium does however require special skills and equipment. Corrosion is less of a problem if the correct grade of material is used and if precautions are taken where steel and aluminium meet, e.g. at coamings. The spacing of stiffeners can be increased in an aluminium structure and this has the added advantage that the number of stiffeners and hence the weight and cost of the structure can be further reduced. The major disadvantage of aluminium is that a suitable grade of material for access equipment costs approximately eight times as much per tonne as mild steel, with the overall effect of increasing the cost of the cover by up to three times.

Aluminium covers have been fitted to deep tanks (Section 5.12.1) because they are so much lighter. Since deep tank covers are often made in one piece, aluminium covers are easier to handle without power assistance. Aluminium single pull weather deck covers have occasionally been fitted in the past, sometimes experimentally (see Fig. 9.21). Aluminium can be used for Ro-Ro

Fig. 9.21. A cargo vessel fitted with aluminium single pull hatch covers.

vessel access equipment but it is usually limited to small ramps and car decks. It is less resistant to fire damage than steel.

9.5.3 Glass reinforced plastic (GRP)

GRP structures are light but the modulus of elasticity of GRP is only 7% of that of mild steel and excessive deflection is a problem which invariably accompanies its use. Additional stiffening is thus necessary, making the structure heavier and more expensive to manufacture. For this reason GRP has not as yet found wide application in the construction of cargo access equipment, although development work is in progress. GRP in 'sandwich' construction could be particularly suitable for covers in refrigerated ships because insulation can be built into them during the manufacturing process at very little extra cost, whereas the present system of insulating steel covers is expensive and time consuming.

9.5.4 Wood

Wooden covers are no longer supplied to weather decks of new cargo ships, although on rare occasions they are fitted in tween-decks as mentioned in Section 5.12.2. Classification Society rules lay down scantlings for beam and board covers because of LLC requirements; in practice they are now mainly for the benefit of repairs to older ships.

9.6 Miscellaneous

9.6.1 Fire regulations

If a large cargo space in any dry cargo ship is closed by means of steel hatch covers having fireproof joints and effective means of closing all ventilators and other openings leading to the holds, it is not necessary to provide a fire-smothering system in the hold, as far as the Classification Societies are concerned. Aluminium and GRP covers however are not acceptable in this regard. In Ro-Ro vessels particular attention must be paid to means of preventing the spread of fire. Access equipment, such as ramp covers and bulkhead doors, is often used for this purpose.

9.6.2 Actuating mechanisms

The drive mechanisms of the various types of access equipment are briefly described in Chapters 5 and 6. The design of winch-operated equipment imposes no special requirements on the ship, other than care when siting winches and sheaves to ensure satisfactory operation. Electric and hydraulic drive systems usually consist of standard units, e.g. 10 kW capacity. In the case of the latter, as many hydraulic units as are required are coupled together to provide the necessary power. Failure of one unit will not prevent operation, but merely slow it down.

Total power requirements rarely exceed 100 kW per ship except for Ro-Ros, so that as operation is only required for brief periods, no great demand is made on the ship's auxiliary power capacity. Reaction forces on the ship structure need to be allowed for by the ship designer. Although such forces are usually small for wire-operated systems, hydraulic operation can impose point loads of 100 tonnes or more, requiring special attention to local structural design.

9.6.3 Ventilators

Bulk carriers usually have ventilators in their hatch covers. These are not normally necessary in general cargo ships having mast house ventilators. Ventilators in bulk carriers intended for the grain trade are often designed to double as grain feeders for topping up the coamings or for loading when the hatches must be closed, e.g. when it is raining.

9.6.4 Operation in Arctic conditions

Access equipment, especially stern ramps and bow visors, is often fitted with small hydraulic jacks which initiate opening when the equipment has become iced up in winter. The jacks have a very small travel and are intended merely to move the access equipment sufficiently to break the ice. For ships operating in arctic climates, special attention must be paid to materials, lubrication and detailed fittings so that equipment can be readily operated.

References

[9.1] Halle, P. OBO Carriers. *MacGregor News*, No. 60, 1971.

[9.2] Lockhart, R. G. Hatch Cover Design and Related Experience. Joint Conference on Hatch Covers, Royal Institution of Naval Architects, Institute of Marine Engineers, Nautical Institute, January 1977.

[9.3] Meek, M. The Structural Design of OCL Container Ships. *Trans. Roy. Inst. of Naval Architects*, 1972.

[9.4] McNicol, M. C. An Operator's Views and Experiences. *Conference papers, Ro-Ro 77*, Business Meetings Ltd., 1977.

CHAPTER TEN

Access Equipment in Service

Summary

In this chapter, the nature of the claims for cargo damage arising from defective access equipment is outlined, together with a general discussion of the incidence of defects and the causes of the most important ones. A broad outline of maintenance requirements and practices is given and some examples of changes in access equipment following ship conversions are briefly described.

10.1 Cargo claims

Goods in transit are always susceptible to damage or loss and provisions for the settlement of claims arising from unsatisfactory cargo out-turn constitute an important element in the overall cost of transporting goods from one place to another. The level of claims made against an individual ship operator may be used as a guide for assessing his insurance premiums and P & I (Protecting and Indemnity) contributions, and excessive numbers of claims may ultimately be paid for in goodwill as well. Thus, irrespective of his obligations as a carrier, the ship operator has strong incentives to ensure that goods are properly looked after while they are on board his vessels.

10.1.1 Seaworthiness

Seaworthiness is the fitness of a ship, its structure, equipment and manning, to undertake a voyage and encounter the ordinary perils of the sea. It is necessary to distinguish between 'statutory seaworthiness' and the implied or express warranty of seaworthiness contained in a contract of affreightment* or a contract of marine insurance. Breach of statute, e.g. failure to have a valid load line certificate, renders a ship unseaworthy and is a criminal offence. It is not the usual basis on which cargo interests make claims

---

* In its legal sense, contract of affreightment means any form of contract for the carriage of inanimate things and/or livestock.

against a ship, however. More important in this respect are the warranties given by the shipowner every time he undertakes to carry cargo.

The interpretation of the principles of maritime law varies from one country to another. At English common law there is an absolute warranty of seaworthiness in contracts of affreightment. This means that at the beginning of every voyage (or every stage of a voyage where it is divided into distinct parts) the shipowner guarantees that his ship is seaworthy and in every way fitted to embark upon it. 'Cargoworthiness' is an essential element of seaworthiness.

The common law position is modified for liners by the Hague Rules (incorporated in the Carriage of Goods by Sea Act) which do away with the absolute undertaking of the carrier to provide a seaworthy ship and instead require him to 'exercise due diligence' to ensure that his ship is seaworthy. In practice this is only slightly less onerous than the absolute undertaking. The Hague Rules do not apply to charter-parties, but it is now customary for these to include a clause which puts chartered vessels on the same footing as liners in respect of their rights and obligations in matters of seaworthiness.

As a general rule, therefore, steps must be taken, at the outset of a voyage, to ensure that all cargo spaces are fit (with regard to cleanliness, temperature, humidity for example) to receive the goods placed in them, and to ensure also, that the goods will be maintained in a satisfactory condition in the face of the ordinary perils of the sea such as storms, rolling and water on deck.

10.1.2 Cargo claims arising from sea water entry

The causes of cargo loss or damage are too numerous to mention in detail, but according to a widely-held belief, an important one is sea water entry through hatches and other access openings. It is not possible to state precisely either the number of claims arising annually from this cause or their total value, since most P & I clubs (who are responsible for settling the majority of such claims) neither publish nor, in many cases, keep the necessary statistics. However, one estimate [10.1] suggests that cargo claims constitute about 35% of the total value of a typical P & I club's business, with sea water entry resulting from access equipment failure accounting for some 15% of this (i.e. 5% of the total).

This estimate is confirmed by further evidence collected by a single insurer [10.2] which shows that sea water entry claims currently represent about 2% of the total received, a proportion that has steadily declined from around 7% in the years 1951–55. It seems reasonable to associate the decreasing relative importance of such claims with the increasing use of steel instead of wood for hatch covers during this period.

Materials such as newsprint, tobacco and cement are hygroscopic and typical of many goods that deteriorate in transit if they are not kept dry. The absorption of even small quantities of water may give rise to substantial claims. In recent years the formation of rust on steel products has also been a frequent source of claims. However, whether or not cargo owners attempt to recover and the magnitude of claims consequent upon their decision to do so, often depend as much on external factors such as the state of the market for their goods at the time of delivery, as on the extent of any physical damage that they have suffered.

It appears that damage arising from the entry of sea water through cargo accesses occurs mostly in bulk carriers and multi-deckers, although precise information on this is unavailable. Leakage around hatch seals, particularly at the cross-joints, is primarily responsible for such damage. This problem is less serious in container ships where each item of cargo is protected by its container, and even when leakage through the hatches occurs, little damage results. Sea water damage is also relatively uncommon in Ro-Ro vessels where stern ramps, around which leakage could occur, are usually sheltered from the severest effects of waves and weather, and bow ramps are protected by visors.

Not all wet damage is caused by salt water. Often, condensation (sweat) resulting from improper ventilation during the voyage, or rain during or prior to loading, is responsible. Thus whenever wet damage is discovered it is necessary to ascertain by chemical test whether fresh water or salt water is present.

The formation of sweat in a ship's hold depends on the moisture content of its cargo and the temperature and relative humidity of the air. It can be prevented by ventilation, but in many modern vessels the arrangements for natural ventilation are inadequate.

10.2 Hatch cover surveys

Regular inspections of hatch covers are carried out by Classification Societies, both to satisfy their own rules and to fulfil their responsibilities as load line assigning authorities. The surveyors undertaking these inspections must be assured that the hatch covers are maintained in good condition and that they remain weathertight when closed. They may require chalk tests and hose tests (see Section 10.2.1) to be performed, although in container ships, these are usually dispensed with at all inspections other than those for four-yearly special surveys or when major structural repairs have been effected.

Hatch cover inspections are also carried out from time to time by cargo surveyors, usually, but not always, acting on behalf of charterers or cargo interests. These inspections are commonly conducted during loading or immediately before departure. Their purpose is to ensure that a particular cargo is properly loaded into the ship and adequately protected against sea water damage by the hatch covers. Hatch surveys are also made before discharging begins in order that the condition of the cargo, and the extent and cause of any wet damage, can be assessed before it is disturbed. Wet damage is often concentrated on the surface of the cargo in patterns which directly mirror the arrangement of hatch joints and seals.

10.2.1 Pre-departure checks for hatch seals

Where the risk of sea water damage exists, it is necessary to ensure that all weather deck hatches can be tightly secured with their seals in good order. The condition of seals can easily be checked by spreading chalk along their compression bars and then closing the hatch. On re-opening the hatch, any ineffective seals will be rendered immediately visible by the absence of chalk marks transferred to the contact surfaces of their gaskets. The results of chalk tests must be treated with caution, however, because while they indicate whether or not compression exists, they do not indicate its degree. Thus a hatch cover could successfully pass a chalk test even though the compression of the rubber gaskets is less than that required to effect a proper seal.

Whenever appropriate (at a cargo surveyor's request for instance) chalk tests should be supplemented by hose tests. These are performed by directing a high pressure jet of water at the joints of a closed hatch. Any water penetrating the seals should be discharged through the drain holes. Hose tests should be carried out before loading has commenced to avoid damage to the cargo. But they can never reproduce the conditions that a vessel will experience at sea and satisfactory test results are no guarantee that hatch seals will remain tight during the ensuing voyage. For this reason some cargo surveyors doubt the value of hose testing.

Poor seals can be quickly and easily remedied by special adhesive tape which is placed over all joints when the hatch is closed and secured for sea. This can never be more than a temporary expedient for a single voyage. Hatch tapes must be applied in dry conditions or adhesion will be unsatisfactory and they will soon be washed away.

10.3 Hatch cover defects

Modern hatch covers are often large and complicated pieces of equipment. They are required to operate continuously in a harsh environment and it is inevitable that defects occur in them from time to time.

In a recently completed survey of weather deck hatch cover defects in cargo vessels, bulk carriers and container ships, reported to Lloyd's Register during the period 1966–75 [10.3], it was found that the incidence of damage defects varied from 3·5 to 11·7 cases per 100 ship-years of service. These findings are set out in Table 10.1.

It is clear that cargo handling accounts for roughly 20% of all defects occurring in the three types of ship studied. The remaining 80% of defects

Table 10.1. Incidence of weather deck hatch cover damage/defects per 100 ship-years service

Ship type	Cause of damage defect		
	Cargo handling	Other causes	Total
General cargo	0·7	2·8	3·5
Bulk carrier	0·9	4·3	5·2
Container ship	2·3	9·4	11·7

Table 10.2. Percentage distribution of defective hatch cover components

Component	Ship type		
	General cargo	Bulk carrier	Container ship
Whole cover	7·4	6·1	—
Plating	27·2	28·8	89·3
Stiffening	2·2	3·0	10·7
Securing arrangements	6·6	3·8	—
Packing, jointing	52·9	50·7	—
Operating mechanisms	3·7	7·6	—
Total	100·0	100·0	100·0

were found to be attributable to a variety of causes including heavy weather and wear and tear. Container ships appear to suffer relatively more (although not necessarily serious) defects than bulk carriers or general cargo ships, probably because they have a large number of single panel covers which must support concentrated loads.

A more detailed analysis of non cargo-handling damage/defects is set out in Table 10.2 [10.3]. Here the affected hatch components are identified for the three types of ship considered. The distribution of defects is remarkably similar for general cargo vessels and bulk carriers and it is significant that packing and jointing account for easily the largest proportion of the total. Faults in the structural plating also stand out for these vessels but to a much smaller extent than for container ships, where they account for practically all of the non cargo-handling defects reported.

Component defects for all three types of ship are broken down by their location along the length of the hull in Table 10.3 [10.3] which suggests that whole cover and plating defects are more likely to occur forward than elsewhere in the ship. This is not altogether surprising since the severest effects of weather and waves are felt forward. On the other hand, most defects in securing arrangements are likely to occur amidships, which is explained, perhaps, by the fact that bending stresses and deflections set up in the hull are generally high in this region. Comparatively few defects in total occur in securing arrangements, however (Table 10.2).

Operating mechanisms account for few of the total number of reported defects. These are apparently concentrated towards the ends of the hull and it has been suggested that pounding and bow motions together with severe propeller–induced vibrations aft are primarily responsible for this.

The position of the bridge appears to have some influence on the incidence of hatch cover defects. In general cargo ships with bridge aft, an average of 3·6 defects per 100 ship-years has been observed whereas in general cargo ships with bridge amidships, the corresponding figure is 1·2. The overall average for all general cargo vessels is 2·8 non cargo–handling defects per 100 ship–years of service.

Table 10.3. Percentage distribution of hatch component defects with respect to position in the ship

Component	Section of ship			
	Forward	Midships	Aft	Total
Whole cover	52·4	23·8	23·8	100·0
Plating	41·1	27·0	31·9	100·0
Stiffening	41·7	25·0	33·3	100·0
Securing arrangements	26·9	50·0	23·1	100·0
Packing, jointing	35·9	34·5	29·6	100·0
Operating mechanisms	26·7	6·6	66·7	100·0

10.4 Some reasons for hatch leakage

Failure of seals and joints is the most common of all the hatch defects occurring in general cargo vessels and bulk carriers, as was pointed out in the previous section. There are numerous reasons for such failures, even assuming that covers and coamings have been installed to the correct tolerances, but often leakage is as much due to the shortcomings of human operators as to the failings of hatch equipment. Thus the majority of common causes of leakage fall into one or other of the following broad categories.

10.4.1 Deformation of hull and hatch covers

The deformation of the hull girder is discussed in Section 9.1.4. As a ship works in a seaway, the relative deflections of its hull structure and hatch covers may be great enough to allow water to penetrate the seals and enter the holds. This is essentially a design problem and whether or not a ship's hatches leak as a result of such deflections depends very much on the allowances that were made for them at the design stage.

There are two basic approaches to this problem. Either the covers are made to flex with the hull or they are made as rigid as possible and highly elastic gaskets are used to absorb the hull deflections.

10.4.2 Accidental deformation of hatches

Hatch covers receive rough treatment in service. Continual careless opening and closing leads to misalignment, localized damage and leaking seals. Careless cargo handling can have similar consequences.

When cargo is worked by derricks rigged in union purchase, coamings and their compression bars often become deeply scored by the repeated chafing of derrick runners (falls). This can be avoided by conscientious winch driving but at many ports it is difficult to persuade dock labour to such diligence. Impact damage to coamings and stowed hatch panels (and also slings of cargo), which frequently leads to hatch leakage, is also due for the most part to poor winch or crane driving.

10.4.3 Improper battening down

Battening down hatches with manual cleats is a long job in medium-sized or large ships with small crews. It must often be done at night, in inclement weather with the ship proceeding to sea, and in such circumstances the crew may not be as attentive to their duties as they should be. Negligence is a major factor in practically every case of improper battening down; cleats left undone, eccentric wheels not turned up, multiple panels incorrectly aligned and cross-joint wedges not hammered up are among the most common forms in which it appears.

Permanent local damage to gaskets, and consequent leakage, may also occur during battening down as a result of obstructions such as small pieces of dunnage, wires and cargo residues that are left on hatch coamings and between panels. In some cases, failure to clear away such obstructions before operating a hatch may cause it to be unshipped.

10.4.4 Permanent setting down of gaskets

There is a commonly held belief that the tighter a cover is secured to its coaming, the better the seal that is made and the smaller the chance of leakage. This view demonstrates a profound misunderstanding of the way in which hatch seals work.

In order to achieve an effective seal the gasket must be compressed but never to the extent that the elastic limit of its material is exceeded. When a hatch cover is designed, the degree of gasket compression that is necessary to prevent the passage of water is determined, taking into account the nature of the gasket material and the acceptable contact pressure. Its cleats, whether automatic or quick-acting manual, are then arranged to provide the correct compression. Although it is possible to obtain greater compression than this, permanent deformation of the gaskets is invariably the result of so doing, especially when the cleats are tightened in an arbitrary manner as often happens during the normal course of battening down. Once gaskets have been permanently set down in this way (or by work hardening) their characteristics are irretrievably changed and their sealing properties lost.

Some types of hatch cover employ highly elastic rubber seals (see Section 9.4.1). These should never be over-tightened since they form part of a sealing arrangement which, besides keeping water out, also allows the cover and the ship's hull to flex independently.

10.4.5 Corrosion

Corrosion is always a problem in ships and it is a common cause of leaking hatches. Unless steps are taken to prevent it, compression bars become corroded, roughened and eventually waste away. Even minor roughening may seriously impair the effectiveness of hatch seals yet it is not uncommon for ships to be sent for their periodic inspections and repairs with their hatch compression bars in very poor condition. This has been recognized as a major cause of leakage by a number of ship operators who now prefer to fit stainless steel compression bars.

Hatch plating is also likely to corrode if it is not adequately protected. In some severe cases holes may appear in it, either as a result of corrosion directly, or as a result of fractures in steel weakened by corrosion.

In combination carriers employed in oil trades, the inner surfaces of hatch cover plating often suffer severe corrosion unless properly protected. Internal corrosion is also often associated with the formation of sweat.

10.4.6 Blocked drainage channels

When sea water penetrates a hatch seal, it is usually collected in drainage channels (see Sections 5.2.6 and 9.4) and discharged clear of the hatchway. If these channels are allowed to become blocked or restricted, water may spill over onto cargo in the hold below. It is essential therefore, that immediately before battening down, compression bars, gaskets and drainage channels are thoroughly swept clean of dirt and accumulated cargo debris.

Drainage channels, particularly those serving cross-joints, are commonly found to be corroded after some years in service.

10.4.7 Seals and cross-joints

Seals that are continually being made and broken by bringing together and separating gaskets and compression bars, are likely to leak eventually, no matter how well they are maintained. For this reason covers with the smallest number of cross-joints, e.g. side rolling, often perform best with regard to leakage. Where such covers are not suitable permanent flexible membranes, such as those linking adjacent panels of Rolltite covers, provide a rather better sealing arrangement than the conventional one.

In the chemical and some other non-marine industries, sealing technology has reached a high state of development. Perhaps some of the pneumatic or hydraulic sealing devices commonly employed there could be adapted for hatch covers and other marine purposes, but the benefits that they might bring, such as reduced leakage, would have to be weighed against their higher initial cost.

Fig. 10.1. The results of neglect.

Fig. 10.2. Severe corrosion of a single pull panel cross-joint. The rubber retaining bar is practically eaten through and the top plate is holed at its outboard edge.

10.5 Hatch cover maintenance

There can be no doubt that inadequate maintenance is a major cause of many hatch cover defects. The marine environment is a harsh one. Damp, salt-laden air, water on deck, dusty, abrasive cargoes all take their toll of a ship's structure and fittings which deteriorate rapidly if proper preventive measures are not taken. Yet it is too often evident that such measures are neglected by the condition of some ships when they are presented for survey and repair.

It is unquestionably more difficult for a crew to carry out extensive main-tenance at sea today than it was in the past when ships were smaller and slower. Crews have not increased in size as larger ships have entered service. Rather, the reverse is true. And not only has the amount of maintenance to be done grown considerably, but with the greater speeds of modern vessels, the crew has less time to do it in. For while ships spend more time at sea nowadays, higher speeds mean shorter passages between ports, and also a greater incidence of spray, even in fine weather, which effectively prevents much deck maintenance. The sheer size of modern hatch covers is another factor that makes their maintenance more difficult, while access to all parts is some-times difficult to arrange.

Thus, with the limited resources normally available on board ships, it may be unreasonable to expect anything other than simple routine maintenance or essential repairs to be carried out at sea. Many ship operators have recog-nized this by making use of the maintenance services now offered by hatch cover manufacturers.

10.5.1 Planned maintenance

Maintenance, whether carried out by the crew or by special maintenance squads, must be properly planned if it is to be effective, and the larger and more complicated the equipment to be maintained, the more important it is that comprehensive records are kept. Traditionally much of the maintenance that has been done on ships has been carried out in an ad hoc manner with individual components being scaled, cleaned, painted or renewed whenever the need becomes apparent. This may be satisfactory when the chief officer, chief engineer or whoever is responsible for maintenance, is familiar with the condition of every single item under his charge, but it is generally an un-acceptable arrangement in large modern vessels where personnel turnover is high. In such circumstances, many items are likely to receive more attention than is their due, while others are likely to be ignored completely.

Planned maintenance is nothing more than maintenance which is organ-ized in an orderly and logical manner to ensure that each component receives the correct amount of attention at the appropriate intervals. Numerous schemes have been devised by equipment manufacturers and ship operators to suit the particular circumstances of different ships, but all have certain features in common, the most important of which is a comprehensive recording system which contains information about every component to be maintained, such as its type, location etc. and also details of its maintenance requirements and history.

Fig. 10.3. Peak-topped cover undergoing reconditioning. New beams are being fitted into a panel which has been removed from a ship to the manufacturer's premises.

Hatch cover manufacturers usually provide comprehensive recommendations and instructions to facilitate setting up a formal planned maintenance scheme, distinguishing between the work that can (and should) be done by the crew and that which should be left to specialist maintenance squads or repair yards. Typical of the work which can be done at sea are the following.

(i) Drive box or electrical compartments – monthly examinations and checks for watertightness.

(ii) Drive chains, trolley, adjusting devices – to be cleaned and greased regularly.

(iii) Seals, compression bars, coamings – to be inspected and cleaned at each port – thoroughly checked and repaired/renewed at regular intervals.

10.5.2 Maintenance facilities provided by manufacturers

The ready availability of supplies of spare parts is an essential prerequisite for effective maintenance. In an international industry like shipping, nothing less than a worldwide network of reliable suppliers is acceptable and in recognition of this, hatch cover manufacturers have established extensive after-sales service facilities. Besides spare parts, these services often include the availability at short notice of a trained maintenance engineer and at major centres, skilled maintenance squads as well.

Because ships nowadays customarily spend only short periods in port, it is becoming increasingly common for maintenance squads to make short sea passages in order to complete work which cannot be finished before departure.

10.5.2.1 *Example of the work of seagoing maintenance squads*

In 1975 seagoing maintenance squads successfully carried out major repairs to the single pull covers fitted in a 7000 tdw cargo vessel built in 1961. The vessel was trading on time charter between the east coast of Canada, the U.S.A. and central South America and her managers opted to have the bulk of the repairs carried out at sea in order to minimize off-hire time. The sequence of events was as follows.

(i) The vessel was surveyed and a full report submitted to her managers.

(ii) Various repair options were suggested, but the phased renewal of panels and their installation by a seagoing squad was accepted as best fitting the managers' requirements.

(iii) Prefabrication of 14 new panels and 18 cross-joint web units was begun in the U.K.

(iv) Prefabricated units were shipped to Halifax ready for the vessel's arrival.

(v) A seagoing maintenance squad was despatched to Halifax to lift ashore 14 existing panels and fit 14 new ones.

(vi) Of those panels brought ashore, 3 were condemned and the remaining 11 were shipped back to the U.K. for repairs.

(vii) The seagoing squad travelled with the ship, working during the passage on chain adjustments etc. and in port fitting the new cross-joint web units on existing panels before returning to the U.K.

(viii) On the vessel's return to Halifax, 11 panels that had been renovated in the U.K. were waiting for her, and were exchanged for a further 11 panels that were beyond repair. The seagoing squad were re-engaged at Halifax for this work and sailed with the ship on her next voyage to Belize. This major job was carried out over a two month period with no loss of charter hire.

This was a more complicated task than seagoing squads are normally called upon to perform – they are more usually employed on re-rubbering and overhauling hydraulic systems – but it demonstrates how large scale maintenance can be carried out without interrupting a vessel's schedule.

10.5.3 General maintenance

10.5.3.1 *Steel work*
Mild steel quickly rusts if it is not properly protected, and once begun the process is not easily arrested. Often it begins before a ship embarks on its maiden voyage.

Manufacturers usually give new hatch covers only a priming coat before dispatch, unless a higher standard of finish is specified by the customer. However primer alone may be insufficient protection for hatch covers stored in open stockyards to await installation and final coating. Generally it is easier to paint covers thoroughly before they are installed and their various fittings are attached. Moreover, many modern coatings, such as epoxy paints, must be applied under controlled conditions and these are more likely to be available in the manufacturer's shop than in an open fitting-out berth. Thus there are good grounds for including painting in the fabrication process.

Paint coverings become damaged in service by impacts and abrasions, and corrosion can then be retarded only be repainting. Ideally this should be carried out in the manner prescribed by the paint manufacturer but in practice this is rarely possible – salt-laden air is not conducive to good paint adhesion! Moreover, some parts of hatch panels, such as cross-joints, may normally be inaccessible during a ship's operation.

The inner surfaces of hatch panels are also difficult to paint, especially in single deck ships. Some operators have experimented with grease paint, which appears to offer good protection for these areas. But grease paint never forms a firm skin and collects dirt and grime, so rendering a vessel unfit for the carriage of grain. For this reason some operators specify double-skinned panels with a rust inhibitor in the void space. This only partly solves the problem, however, because the outer surface of the lower skin must still be painted.

Care should be taken to avoid painting over and blocking drain holes.

10.5.3.2 *Seals*
The principal defects that occur in seals are outlined in Section 10.4.

Standard rubber gaskets can be expected to last from four to five years of normal service. Their life is severely curtailed by over-compression and contact with abrasive materials. Where local damage to gaskets occurs, a common practice is to glue additional layers of rubber over the affected area. This practice is seldom effective and is to be deplored. Any section of damaged gasket must be renewed.

In freezing conditions special grease or commercial glycerine should be spread over the surface of all gaskets in order to prevent them sticking to their compression bars.

The roughening of mild steel compression bars can be alleviated to some extent by buffing. Where their contact edges are severely wasted, however, it may be necessary to build them up with weld or even renew them completely.

10.5.3.3 *Cleats*

Quick-acting cleats are probably the most common and most frequently handled of all hatch cover components. They exert the correct degree of compression through thick neoprene washers which, in time, tend to lose their elasticity. When this happens the cleat must either be adjusted or, more often, replaced. Galvanized cleats have been found to have a longer life than ordinary steel cleats.

In order to cover replacements, ships should carry a good stock of spare cleats. Some operators recommend an additional 25% of the total cleats in use as spares.

10.5.3.4 *Chains*

On single pull covers, the connecting chains or wires between individual hatch panels must be inspected for signs of stretching. When the design tolerances are exceeded, and the chains on one side are not the same length as those on the other side, the panels tend to turn or 'crab' as they are pulled forward. In serious cases, panels may jam or become unshipped.

10.5.3.5 *Hydraulic systems*

Dust and fine particles mainly from cargo sources, are amongst the principal enemies of hydraulic systems. They settle on lifting and hinge jacks during loading and discharging and lodge in their seals which eventually fail unless they are regularly inspected and cleaned.

The corrosion of steel hydraulic pipes is another common source of failures. Often this is due to inadequate paint protection (enhanced by the siting of hydraulic pipes too close to flat surfaces so that it is possible to paint only those areas that are immediately visible). Because of the internal pressures that hydraulic pipes must withstand, they have thick walls and it usually takes some years before they are sufficiently wasted to fail. However, when failure does occur it is invariably necessary to replace complete sections of pipe.

10.6 Hatch cover maintenance costs

The precise cost of maintaining steel hatch covers is difficult to determine and can vary widely between different ships in different trades. Many ship operators simply include expenditure on hatch cover maintenance along with other general deck maintenance and do not include work done by the ship's crew. This could be taken as an indication that they do not consider hatch cover maintenance a particularly important cost item. But it is probably an over-optimistic view.

It is clear, however, that annual maintenance costs depend primarily on the type of equipment installed, its age and the care with which it is treated. Thus in a comparatively new ship, for example, the annual cost of maintaining access equipment might be as little as 1–2% of its replacement price, while for an elderly ship in a demanding trade, this percentage could be nearer 5–10%.

It is unwise to assume that the total cost of any item of access equipment over its life can be gauged from its initial cost alone. The cheapest initially is not necessarily the cheapest in the long run and the savings in maintenance costs which may be achieved by, say, fitting stainless instead of mild steel compression bars may well outweigh their higher initial cost (see Section 7.3.2).

10.7 Ro-Ro access equipment maintenance

As pointed out in Section 10.1.2 there have been comparatively few major claims for damage to cargo arising from the failure of Ro-Ro access equipment. This may be taken as an indication of the reliability of such equipment in service. Nonetheless, the consequences of a major failure could conceivably be very serious and it is essential that the maintenance of ramps, doors and elevators is not neglected. It is possible that the good record of Ro-Ro vessels is in some measure due to the high standards set by their operators, many of whom can be counted among the world's leading shipping companies.

It is not usually possible for the crews of Ro-Ro vessels to undertake anything but the simplest maintenance (such as routine greasing) of access equipment, both because of its complexity and because it is rarely accessible at sea, while port turnround times are short. Large items of equipment, such as quarter ramps, may have automatic lubricating and other facilities built in. Manufacturers also offer extensive after-sales service facilities for Ro-Ro equipment and are usually prepared to undertake regular maintenance supervised by their own engineers. Such work may be carried out under continuous Classification Society survey. The same general principles as described earlier for hatch covers apply.

10.8 Ship conversion

10.8.1 General considerations

It is not unusual for merchant ships to undergo substantial structural alterations at some stage in their working life, either to improve their capacity to perform their original function in changed circumstances, or to adapt them to a completely new purpose. This practice has a long history, extending to ancient times. But like so many aspects of shipping, it has become more highly developed and more important since the end of the Second World War. The variety of conversion work that has been undertaken is very wide. It includes the lengthening of war-built *Liberty* vessels to increase their deadweight, the modification of numerous tankers and general cargo vessels to create the first generation of cellular container ships, and the conversion of passenger vessels into cruise liners, school ships and temporary accommodation for construction site personnel.

The changes which the shipping industry has experienced in recent years, have occurred so rapidly that many ships have become unsuitable for the tasks for which they were built long before their expected lives have expired. This is particularly true of tankers, for instance, whose average life expectancy decreased substantially during the 1960s. In some cases it has even extended to new vessels which, in the often long lead interval between their initial ordering and delivery, have been rendered unsuitable for the employment for which they were originally intended. At the same time new-building prices have risen dramatically. All these factors together have led to the creation of a ready market for sound secondhand tonnage suitable for conversion.

Most ship conversions can be completed in a short time compared with newbuilding, which means that ship operators can sometimes decide upon conversion as a tactical response to a particular market situation. Common reasons for undertaking conversions are:

 (i) to improve operational flexibility – e.g. adapting a bulk carrier for the occasional carriage of cars;
 (ii) to adjust the overall structure of a fleet – where a fleet has become too dependent on a particular market sector which shows signs of becoming depressed;
 (iii) to improve vessel performance – often fairly straightforward modifications such as re-engining, lengthening or changing from wood to steel hatch covers in order to make a vessel better able to compete against new ships entering service;
 (iv) to raise a vessel's resale value – this may be worth considering if the potential resale value is greater than the current value plus the cost of conversion.

There are three basic types of conversion which can be summarized as follows:

 (i) Other use: changing vessel type
 (ii) Same use: 'jumboizing' (increasing cargo capacity by lengthening or deepening)

Table 10.4 Conversion options and benefits

Type of Conversion	Market Condition	Benefits of Conversion
Other use (change of type)	Oversupply or poor employment potential in one sector of operation Changing trade structure by volume, form or route New company ventures	Redeployment within company Resale at profit under depressed market conditions
Same use (jumboizing)	Increasing demand in existing trade Poor operating cost/performance ratio	Increased capacity on route Improved fuel effectiveness
Same use (modernization of equipment)	High cargo handling costs Competitor introduces faster or more efficient tonnage	Reduced turnround time in port Meeting competition from modern tonnage

(iii) Same use: modernization (e.g. fitting Ro-Ro access equipment for multi-purpose operation).

Examples of their possible applications are given in Table 10.4 [10.4].

10.8.2 Examples of ship conversions

The following examples demonstrate the wide variety of the many conversions that have been undertaken in recent years, and the important contribution that modern access equipment has made towards their success.

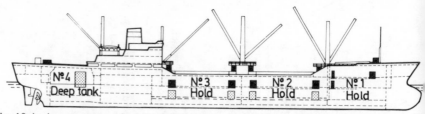

Fig. 10.4. Arrangement of additional lower tween-deck, twin hatches and deep tank aft installed in the 13 500 tdw *Aludra* on changing ownership during construction.

Fig. 10.5. Conversion of a 7 000 tdw cargo ship into a cellular container ship. Note the arrangement of cell guides and the forward end coaming. A pontoon cover is being lifted into place.

10.8.2.1 *Installation of extra decks in general cargo tonnage*

The standard 13 500 tdw cargo vessel illustrated in Fig. 10.4 was fitted with an additional tween deck on change of ownership, after it had reached an advanced stage of construction. The deck, together with two sets of twin

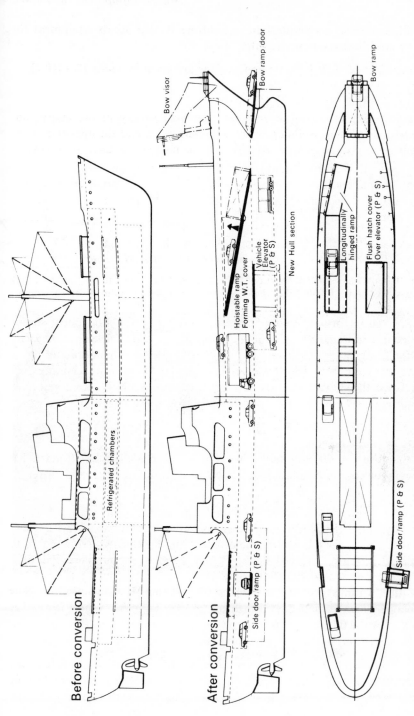

Before conversion

Refrigerated chambers

After conversion

Side door ramp (P & S)

Bow visor

Bow ramp door

Hoistable ramp.
Forming W.T. cover

Vehicle
Elevator
(P & S)

New Hull section

Bow ramp

Longitudinally
hinged ramp

Flush hatch cover
Over elevator (P & S)

Side door/ramp (P & S)

Side door

Fig. 10.6. Conversion of a small general cargo vessel into a pallet carrier. See also Fig. 2.5, which shows the same vessel.

hatches, was fitted in Nos. 2 and 3 holds in 35 days. At the same time, oil-tight side hinged covers were fitted in No. 4 tween-deck and No. 4 lower hold was converted into deep tanks for the carriage of vegetable oil.

10.8.2.2 *Conversion of general cargo vessels into container ships*

Many of the first generation of container ships were converted general cargo vessels and tankers and even today conversions of this sort are still under-taken. In the simplest cases, general cargo/container ship conversions involve the removal of cargo handling gear, the enlarging of hatchways to standard container dimensions and the fitting of cell guides and pontoon covers (see Fig. 10.5). Often, additional wing ballast tanks are also necessary new additions.

10.8.2.3 *Conversion of general cargo vessel into a Ro-Ro/pallet carrier*

In this case a conventional, ten year old, 2000 tdw cargo vessel was converted to accept Ro-Ro cargo and pallets via side doors and a bow ramp, permitting cargo to be handled at a combined rate of some 200 tonnes per hour.

The entire forward end of the hull was replaced with a new section which incorporated a bow visor and ramp, an internal vehicle deck, elevators to the lower hold, and a ramp to the weather deck. Side doors and ramps were fitted on each side of the original after hold. The conversion, which increased the ship's length by 15 metres, its steelweight by 350 tonnes and deadweight by 500 tonnes, is illustrated in Fig. 10.6.

10.8.2.4 *Conversion of a tanker into an OBO carrier*

In this case a 42 000 tdw tanker was converted into a combination carrier. This involved the installation of 11 sets of twin side rolling hatch covers.

10.8.2.5 *Conversion of a tanker into a Ro-Ro trailer carrier*

An 18 year old 18 000 tdw tanker was recently converted into a Ro-Ro trailer carrier. The upper portions of the original tank bulkheads were removed to make way for a vehicle deck having sufficient space for 50 trailers. This deck can be connected to the shore by a bow ramp protected by a visor. An additional ramp provides access to the weather deck where 150 20 ft containers can be loaded, two-high on trailers. To make room for weather deck stowage the bridge structure, which was originally in the traditional tanker position forward of amidships, was removed and resited adjacent to the after deck house. This novel conversion is illustrated in Fig. 10.7.

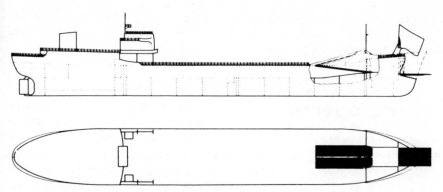

Fig. 10.7. Conversion of the 18 000 tdw tanker *Marilen* into a Ro-Ro trailer carrier.

References

[10.1] Goldie, C. W. H. The views of a P & I Club manager. Joint Conference on Hatch Covers, Royal Institution of Naval Architects, Institute of Marine Engineers, Nautical Institute, January 1977.

[10.2] *Ports of the World. A Guide to Cargo Loss Control.* Insurance Corporation of North America.

[10.3] Lockhart, R. G. Hatch Cover Design and Related Experience. Joint Conference on Hatch Covers, Royal Institution of Naval Architects, Institute of Marine Engineers, Nautical Institute, January 1977.

[10.4] *Ship Conversion: Adapting to a Changing Market.* International MacGregor Organisation, 1977.

CHAPTER ELEVEN

Recent Developments and Prospects

Summary

The aim of this last chapter is to discuss recent developments in ship types and then to indulge in some 'crystal gazing', to visualize changing patterns of trade in dry cargo and the consequent impact on cargo access equipment. Prediction of future trends can be readily upset by international developments in the world movement of commodities, but foresight brings its own rewards. While it is stimulating to consider the distant future, it is foolhardy to be unaware of more immediate potential developments.

11.1 Recent developments

Before discussing possible developments and the impact of constraints upon future ship design and associated equipment, it is pertinent to review some of the developments in existing ship types. Experience with these is all part of the pattern of gradual change, a continuation of the process of natural selection in technology.

11.1.1 Ro-Ro

The adoption of Ro-Ro systems with their abundant configurations for external doors and internal ramps has been widespread, as shown in current trade journals, ships on order and reports of recent conferences. Now, in the late 1970s, Ro-Ro is clearly in its heyday with new services and new equipment coming into effect each month, while designers try to keep abreast of ship operators' requirements.

The introduction of a Ro-Ro service satisfies the short-term demand for an expediency to avoid congestion in ports where cargo is handled traditionally using cranes or derricks; but the capital expenditure is high and the hold space utilization relatively poor, which suggests that on deep sea routes the boom may be short-lived once ports are able to re-establish an orderly demurrage-free throughput. However, should more ports prepare their own equipment for end-on berthing in conjunction with adjustable or portable floating bridge ramps, then new but simpler Ro-Ro ships could be built or converted. The question remains as to who will provide the investment, and where. The trend in World Bank loans has been to favour internal transport systems rather than ships. Furthermore, port authorities are discussing whether the design of Ro-Ro berths and ships could be harmonized, which would be helpful to access equipment designers in the same way that the standard container was to ship designers.

11.1.2 Hatch covers

Less spectacular have been the developments in hatch covers. Hatch sizes continue to increase, still with demands that they be able to support deck cargo. Retaining the ability to be independent of port facilities, more bulk carriers are being fitted with cranes and larger hatches to give direct vertical access to a greater proportion of the hold.

Inevitably hatch cover weights increase with consequently greater loadings on ship structure and on power requirements for the actuating machinery. Improvements are continually being made to provide a higher strength-to-weight ratio, easier maintenance and standard fittings.

The hatch cover itself could play a more useful part in cargo handling work and there are examples of use as a mini-barge and as a stability pontoon for aiding heavy lifting operations. Progress with automation in operating covers continues, but attention always needs to be given to improving stowage arrangements and reducing the space consumed. At times hatch covering seems an ungainly procedure ripe for new thinking on simpler, yet still effective, covering and closing appliances.

11.1.3 New ships – BORO and barge carriers

New ship types have been created and put into service, while at the same time progress on standardizing existing forms continues. The shipbuilder and access equipment manufacturer may welcome a steady state since diversification introduces individual problems which only occasionally become the forerunner of a production series. Outstanding examples of new ships are a bulk/Ro-Ro combination carrier, termed BORO, and barge-carrying vessels.

The BORO design shown in Fig. 11.1 has an unusual, almost traingular, midship section which provides good stability characteristics without resorting to the carriage of substantial ballast capacity. Two 11 000 tdw ships recently built to this concept carry cars and trailers on decks above oil cargo tanks, thereby avoiding the problems of cargo access and the dangers of reduced subdivision below the main deck. A further design Bulkliner-Tankliner 2000 allows the carriage of Ro-Ro freight above dry bulk cargoes which are handled by a conveyor belt system. Bow and stern ramps only are required and there need be no hatches in the decks.

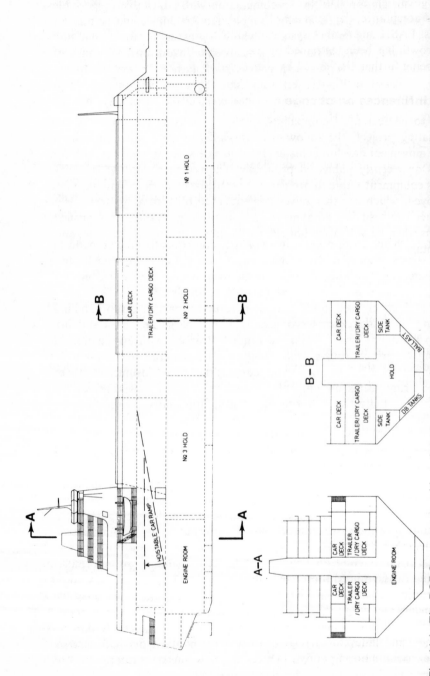

Fig. 11.1. The **BORO** concept provides a large amount of reserve buoyancy and enables the separate carriage of dry bulk cargoes handled conventionally through hatchways and cars and trailers moving across a stern and an internal ramp. The drawing shows the Bulkliner version of 11 000 tdw. Note particularly the triangular shape of the midship section which obviates the need for substantial ballast capacity.

Barge carriers of the types described in Section 8.4.2 have not gained widespread acceptance; the units are often too large with deadweights up to 850 tonnes. LASH and Seabee systems operate mainly from the United States, but growth has been restrained by the adherence to containerization and is now being further challenged by the spread of Ro-Ro systems.

11.2 Influences on change

Privy to no secrets of companies perhaps already in the throes of some fascinating project, the following paragraphs provide some thoughts on cargo movement developments likely in the years ahead, together with comments on recently published proposals. The types of ship and associated access equipment required are dependent on the cargo handling techniques employed, which are themselves consequential on trading patterns. If the designer can visualize equipment opportunities and conceive appropriate designs, then he will place both himself and his company at a competitive advantage.

11.2.1 Patterns of change

Ships will continue to be used for cargo transportation on account of the large quantities involved, geographical locations and the economics in operation. Ship types will be modified to suit new trading patterns, to meet new regulations and to reflect changes and improvements in technology which can be used in the marine environment. Cargo forms are likely to undergo some change, for example more slurried bulk cargoes, and this could have a considerable impact on ship designs, as shown on the container ships of the late 1960s. World trade must continue to expand into regions hitherto unexploited or under-populated. Techniques for handling that trade will be assessed to ensure the lowest transport costs represented by a combination of time, convenience and safety.

11.2.2 Trade

Major influences on cargo handling techniques are likely to come from exploitation of the vast resources buried inside the Arctic. Other regions of considerable resource wealth include remoter parts of South America and mineral deposits lying on the seabed in water depths of around 5000 m. Looking further ahead there are plans for man-made islands and undersea communities; in the latter event completely new thinking on cargo handling and underwater cargo access will be required. Offshore loading at exposed locations is already practised by oil tankers, but if big ships are to be used in remote and undeveloped regions, then there is scope for dry bulk handling at single point mooring buoys, possibly involving aerial ropeways for cargo transfer.

11.2.3 Environment

Trade in the Arctic, with its mineral resources but severe environment, is beginning and an increasing number of larger ships are built with ice strengthening to the hull. Using this as an example of environmental impact on the design of ancillary equipment, mechanisms must continue to function in extremes of temperature, avoiding ingestion of ice into moving parts and the incidence of brittle fracture in materials. Operators must be well shielded from the weather while anything more than minor repairs becomes impossible. Equipment reliability becomes paramount in conjunction with simplicity of design and repair by replacement. Cargoes may need to be protected during the loading phase under inclement weather and this could extend opportunities for self-loading and self-discharging conveyor ships, already common in the Great Lakes. As in some existing bulk and combination carriers, one of the hatch covers may need to be suitable for helicopter operations which help to guide ships through ice floes. As Arctic trade develops, imports will be required to sustain the base operation and local welfare. Techniques for handling what would be break-bulk commodities, possibly by palletization, will affect ways in which access from the ship will be achieved in such a cold environment.

11.2.4 Cargo form

The rapid change from break-bulk transport to unitization substantially affected the type of ship used, the method of cargo handling and the design of hatch covers, doors and ramps. Hatch covers no longer needed to be of the multiple panel format, they could be large, simple pontoons, yet strong enough to carry containers stacked three or four high. The size of unit has increased from the 1-tonne pallet to the 20-tonne container and thence to the 460-tonne LASH barges and massive 1020-tonne Seabee barge. The barges may be loaded with break-bulk commodities or containers but still require their own closing appliances. A more modest grouping is four to eight 20 tonne containers, as in the LUF system, which may be handled as a wheeled Ro-Ro unit. The weight and dimensions of such units influence hatchway and door sizes, ramp capacities, handling appliances required and the options for cargo movement whether it be horizontally over ramps or vertically through hatch openings.

Another example of change in the way that cargo is shipped is the existence of special tankage aboard chemical and parcels carriers rather than break-bulk cargo liners which no longer need to have liquid-tight hatch lids, deep tank coatings and heating coils. There is more pre-processing at source of bulk materials to increase the value added and the potential for higher freight rates. Furthermore, some semi-processed bulk materials which are carried either in small bagged quantities aboard cargo liners or in adapted bulk carriers, may next undergo change into a slurry form. Slurries can be readily pumped from shore to ship, even when it is moored remotely at a loading buoy. The ship configuration may be merely of tanker form with only small access hatches and no handling gear. Many of the problems associated with dirty and dusty cargoes can be eliminated by changing to conveyor belt or pumpable systems, in which case large hatches are no longer necessary for ease of access by grab or even by overhead gantry conveyor as shown in Fig. 11.2. Rapid changes in cargo forms are unlikely to occur, but more shipowners are providing specialized vessels in the quest for efficient transport of semi-processed materials.

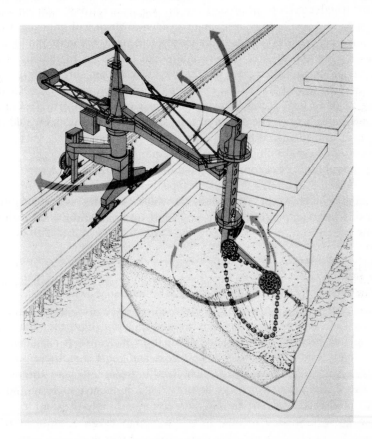

Fig. 11.2. An example of equipment for the continuous discharge of bulk cargoes is this quayside gantry proposed by Paceco Inc. Access can be readily achieved into the hold beneath the overhanging deck plating so that the very large openings needed for rapid discharge by grab are no longer required. The cargo can be reclaimed by means of buckets, screws, suction pipes or digging wheels which feed on to conveyor belts leading ashore.

11.2.5 Regulations

Regulatory requirements can have a marked effect on ship design and hence cargo handling techniques and methods of access to cargo spaces. As mentioned in Chapter 4, most regulations were drawn up with 'conventional' ships in mind, and may not cover the requirements of vessels like Ro-Ros adequately, for instance in respect of bulkheads.

Non-passenger carrying Ro-Ro ships generally have as few bulkheads as possible so as not to restrict horizontal movement. Only small freeboards are required so that the main vehicle deck can accordingly be close to the waterline. The result is an extensive and relatively unimpeded underdeck volume which causes some anxiety on safety considerations. Accidents have happened and regulations are being considered which will improve reserve buoyancy, reduce loss of stability caused by free surface effects, and limit the spread of fire. These may involve additional freeboard to the sill of any access opening and some extra bulkheads. The effect on cargo access equipment may be to restrict the use of internal ramps and encourage the wider adoption of vertical access by elevator from a higher platform deck into a series of holds below. The design of ramps, side and stern doors would need to be re-examined and perhaps modified to meet new regulations, yet another example of external influences on access equipment.

The possible effect of tonnage regulations on the size of paragraph vessels was discussed in Section 4.2.3, but the type of access equipment is unlikely to be altered here. However, astute designers will look for ways of making cargo spaces as large as possible within any limits (or paragraphs) set by regulations.

11.3 Ship types

The designer of access equipment must be aware of trade developments for which new ship types may be designed. New trades may call for re-appraisal of existing cargo handling techniques and equipment. The ship is the vital link between the origin and destination of a cargo. It must carry the cargo safely and efficiently at lowest cost and must match the facilities available, or to be created, at each handling place. Where there is an established cargo flow which can be treated in a homogenous form, the best opportunities arise for purpose-built vessels designed into the total transport system.

11.3.1 Tug-barges

One system that is gaining support as an alternative to conventional tonnage in particular situations is the tug-barge configuration. It has proved satisfactory in U.S. coastal operations with bulk cargoes, even over long distances. Transocean passages have yet to be implemented on a regular basis, although several schemes have been investigated by shipping companies.

An interesting development in the general cargo trade has been the two Ro-Ro barges put into service in 1977, between ports in the South of France and the Red Sea, resembling floating multi-storey car parks (Fig. 11.3). Each can carry 261 trailers on three decks divided into 10 lanes. The barges are moored end-on to the quayside and the twin-lane ramp can be moved along the quay and elevated to one of the required positions. Special tugs were built for towing the dumb barges at about 11 knots.

Fig. 11.3. The floating multi-storey trailer park *Arab Hawk* is shown loading at Fos in the South of France, prior to being towed to the Red Sea port of Yanbu via the Suez Canal. This Ro-Ro development enables end-on berthing of the barge to the quay and access using a portable twin-lane ramp. The dumb barge has a length of 127 m and carries up to 261 12 m trailers at a towed speed of 11 knots.

Operational advantages may eventually ensure their wider acceptance regardless of any technical shortcomings that they may have, for example the reliability of the connection between the tug and the barge. Because a barge or train of barges with their modest dimensions can be taken closer to sites of industrial activity and prepared berths with their attendant facilities, advanced access handling equipment could become redundant in some cases. Cargo handling systems are designed around the barge unit. If the barge is used partly as a temporary warehouse, urgency of discharge or of loading is avoided, thereby eliminating perhaps the need for extensive facilities. Any wider adoption of tug-barge systems in the general cargo trade would undoubtedly affect market trends towards Ro-Ro access equipment, but compensations could be seized by designers aware of prospects and changes in trade patterns and cargo handling techniques.

Fig. 11.4. An example of barge-carrying vessels in *Bacat 1*, built for short-sea trading. On deck up to ten 140-tonne deadweight barges are carried, raised from the water tunnel by an elevator working between the semi-catamaran hulls. A further three 460-tonne LASH barges may be secured in this tunnel which is sealed with a special stern closer. The ship has a length of 103 m and speed of 13 knots.

11.3.2 Barge carriers

A variation on the barge theme involves floating in and out (Flo-Flo) of a mother craft, a concept already used for military purposes and the revitalized 2910 tdw BACAT (Barge Aboard CATamaran), now trading in the Indian Ocean instead of the originally intended North Sea (see Fig. 11.4). A number of design proposals have been put forward including the Capricorn (see Fig. 11.5) which has a bow door allowing LASH or even Seabee barges to be floated in and out of a flooded hold, a stern ramp for Ro-Ro cargoes and a deck gantry crane for handling containers. It is a versatile craft which the designers suggest would be particularly useful on trades between industrialized countries and less developed areas without

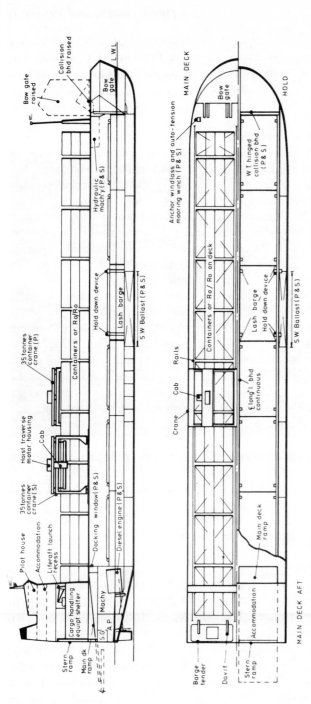

Fig. 11.5. The Capricorn Carrier (type LR-12) is designed as a multi-role vessel. Trailer cargoes are driven aboard up to the main deck via the stern ramp. Alternatively, the main deck can carry 375 20 ft containers stacked by one of the two deck gantry cranes. Additional cargo is stowed in the space below the main deck, which may be flooded to enable 12 LASH barges to be pushed in through the raised bow gate. This version has a length of 146 m and a speed of 15 knots.

deep water ports equipped with cranes or extensive transport facilities. The only access features are bow and stern doors. Providing adequate strength and watertightness is achieved at the bow, more widespread use of this access may be worthy of further consideration.

11.3.3 Bulk carriers

Opportunities for cargo handling and access development in bulk carriers seem most likely in new roles such as the loading and transfer of minerals from seabed mine sites and the transhipment between carrier and production unit while at sea. Such operations are already practised in the tanker trade, and for stores replenishment of warships while at sea.

Conveyor systems may be more frequently fitted, using belts, chutes, air-slides and pneumatic lifts which must be designed to suit the materials expected to be handled. Access into holds need not be conventionally 'over the top'; there is scope for handling through the ends of the ship or through its sides, depending on the nature of the operation.

Bottom dumping ships are presently used in the removal of silt from harbours and estuaries. Spoil is loaded into the hold from above by grabs or from pipes. Rapid discharge is effected through the bottom either by opening doors or by the ship hinging open like an enormous clamshell. Bulk carriers might incorporate this method and discharge onto specially prepared underwater reception areas, provided the cargo was neither affected by water nor washed away by currents. The ship would be speedily discharged and the receiver could recover the material as required using shore-based grabs or a conveyor system, a scheme perhaps suitable in the transport of seabed minerals.

Bulk carriers could transport small quantities of higher value semi-processed cargoes, but since losses occurring during grab discharge would not be tolerated, new ways of handling would have to be found. More attention would have to be given to adequate protection during handling and transport if the cargo were susceptible to moisture or dusty or could contaminate adjacent cargoes or the environment.

11.3.4 Future ship types

Fundamental changes in the present methods for handling and transporting cargo seem unlikely in the next decade or two. Modifications to existing ships and new variations will be tried. The escalation of fuel costs has meant that high-speed ships are harder to justify for the carriage of commercial cargoes except for special services, for example the 30-knot gas turbine passenger/vehicle ferry *Finnjet*. Multi-hull craft like catamarans have good stability characteristics but small deadweight potential and so may be restricted to craneships or special ships requiring large deck areas.

Hovercraft and hydrofoils will continue to be developed for ferry routes, but their payload and range characteristics severely limit their potential for deep sea cargo transport. Cargo-carrying submarines have often been proposed, especially for transport under Arctic ice, but practical problems are very great. Current designs are most suitable for liquid cargoes, partly because large accesses for dry cargo in a pressure resistant hull are difficult to arrange.

It is the way in which existing practices can be put together to achieve lower total transport cost that developments will be sought.

11.4 Materials

The basic material for access equipment is likely to remain steel except in a few special instances as discussed in Section 9.5.

Certain alloys and synthetic materials are used in access equipment fittings to improve durability under marine conditions. With decreasing crew numbers on ships, maintenance work usually has to be left until port where repair or replacement must be swift and effective. Careful attention should be made during specification and design and in the choice of materials to ensure that items of equipment can work effectively and compatibly, be readily overhauled and, when necessary, be quickly replaced.

Where there is a flooding or fire risk, it is likely that steels of various qualities will continue to be used, but in other cases the adoption of aluminium, GRP, or GRP reinforced with steel could provide lighter structures for hatch covers, ramps and doors where weight saving is important.

11.5 Innovation

This book has tried to provide a comprehensive insight into the options available to the ship operator, ship designer and cargo access equipment designer, but the impetus for innovation within transport systems will continue to remain primarily with ship operators and those with cargo interests.

The equipment designer has an important role to play in collaboration with all participants in order to provide the most suitable equipment. Undoubtedly lively discussions will ensue, for example, on the merits of elevators versus ramps, and the designer will be asked to provide access equipment for some quite novel schemes. Here, then, are his two principal contributions: to provide more efficient access equipment for the ships of today and to be ready with innovative proposals for the ships of tomorrow.

Index

References in italic indicate illustrations.